Stucki Side Bearing

Free-swiveling truck—Increased economy and safety
High grade materials—Practically indestructible

IN building the Pyramids, the Egyptians used rollers to move monstrous stones. To move heavy machinery in modern shops, the workmen use rollers to avoid the heavy drag; and 50 or more years ago progressive railroad managers suggested rollers for side bearings, to assist the truck in swiveling freely under the car in passing curves.

In doing so the pressure against the rail in deflecting the truck is very much reduced and the saving therefrom is fivefold, namely, in:

Drawbar Pull — Rail Wear
Wheel Flange Wear — Derailments
Paying Load

In this every railroad agrees, even the few who still use ordinary friction blocks. Among these few, some are now using an expensive, delicate system of oiling the rails to overcome the friction between the sliding surfaces of the side bearing.

Smaller Drawbar Pull

The free swiveling of trucks equipped with roller side bearings allows them to align properly with the tracks without undue resistance which means that the train pulls easier and the locomotive will require less fuel.

Saving in Rails

The saving in rails on curves is simply astounding. This was established by comparing the removal periods on many up-to-date roads with one which yet uses common friction blocks.

Saving in Wheels

The saving in wheels is naturally similar to that on the rails, because the pressure is the same on both.

Fewer Derailments

No doubt the most important saving is that of reducing the dangers of derailments. When the contour of the rail and wheel is worn and the truck is forced to swivel on a curve, the wheel may find it easier to climb the rail than to push the truck around and if the friction at the side bearing is not removed, a costly derailment may be the result.

Larger Paying Load

A reduction in train resistance means that a locomotive can pull more cars. The amount can very easily be determined by tests, road or laboratory. The Stucki Side Bearing showed it to be 10.4 per cent.

Because the underlying principle of a frictionless side bearing is as old as the Pyramids, hundreds of various designs of side bearings were introduced, tried and abandoned in years gone by. Most of them failed on account of being too complicated and many of them on account of using inferior material.

Operating and Construction Features

The A. Stucki Company, Oliver Building, Pittsburgh,, Pa., realizing all that, started about thirty years ago with a two-piece frictionless side bearing, as shown in Fig. 14.120, and it is yet used in connection with special constructions.

About eighteen years ago the Company succeeded in producing the same bearing in rolled steel, as shown in Fig. 14.121. Using a high grade material makes it actually indestructible in service and reduces its repair to practically nothing.

This being a truck side bearing makes it immune to abnormal side bearing clearance.

The cage is made slightly harder than the roller to minimize wear at its bottom.

Stucki Side Bearing, All-Rolled-Steel Construction

For the same reason the bottom in the center is flat for 1½ in. to avoid impinging it on one and the same line.

The roller itself will not wear flat, as it turns more or less every time the train is stopped, presenting constantly new lines of contact.

The one-roller Stucki Side Bearing is made to suit the A. A. R. standard car constructions and takes care of all equipment up to 100-ton capacity, and in length up to 65-ft. mill type cars. This means a tremendous simplification in the storehouses and repair shops, as one side bearing is all that is necessary to keep in stock.

In exceptionally high capacity equipment, such as 22,000-gallon tenders, two rollers are used instead of one as shown in Fig. 14.122.

In passenger equipment and especially in high speed streamline cars, where weights are kept down as low as possible, the side bearing as shown in Fig. 14.123 is used, it being light, low, and dependable.

All inquiries should be addressed to the address shown below.

A. STUCKI COMPANY, Oliver Building, PITTSBURGH, PA.

Fig. 14.124—Woods 7-in. Travel Gravity Roller Side Bearing.

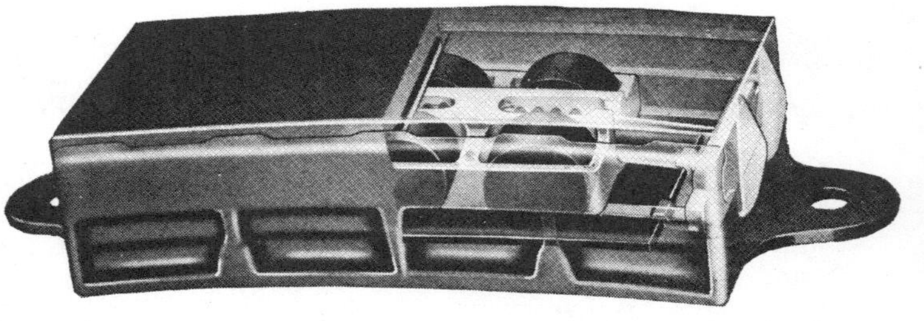

Fig. 14.125—Woods Spring-Controlled Truck Side Bearing for freight and passenger cars.

Fig. 14.126—Woods Counterbalance Truck Bolster Side Bearing for freight cars.

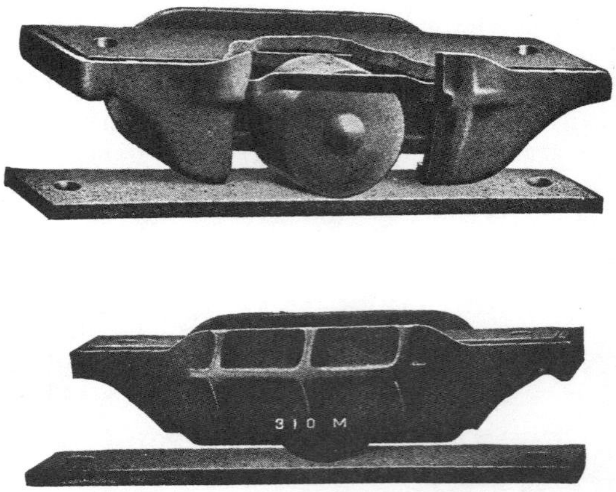

Fig. 14.127—Woods passenger car Side Bearing showing roller and springs.

Edwin S. Woods & Company
Division of Pettibone Mullikin Corp.
(*See also Page* 1133)

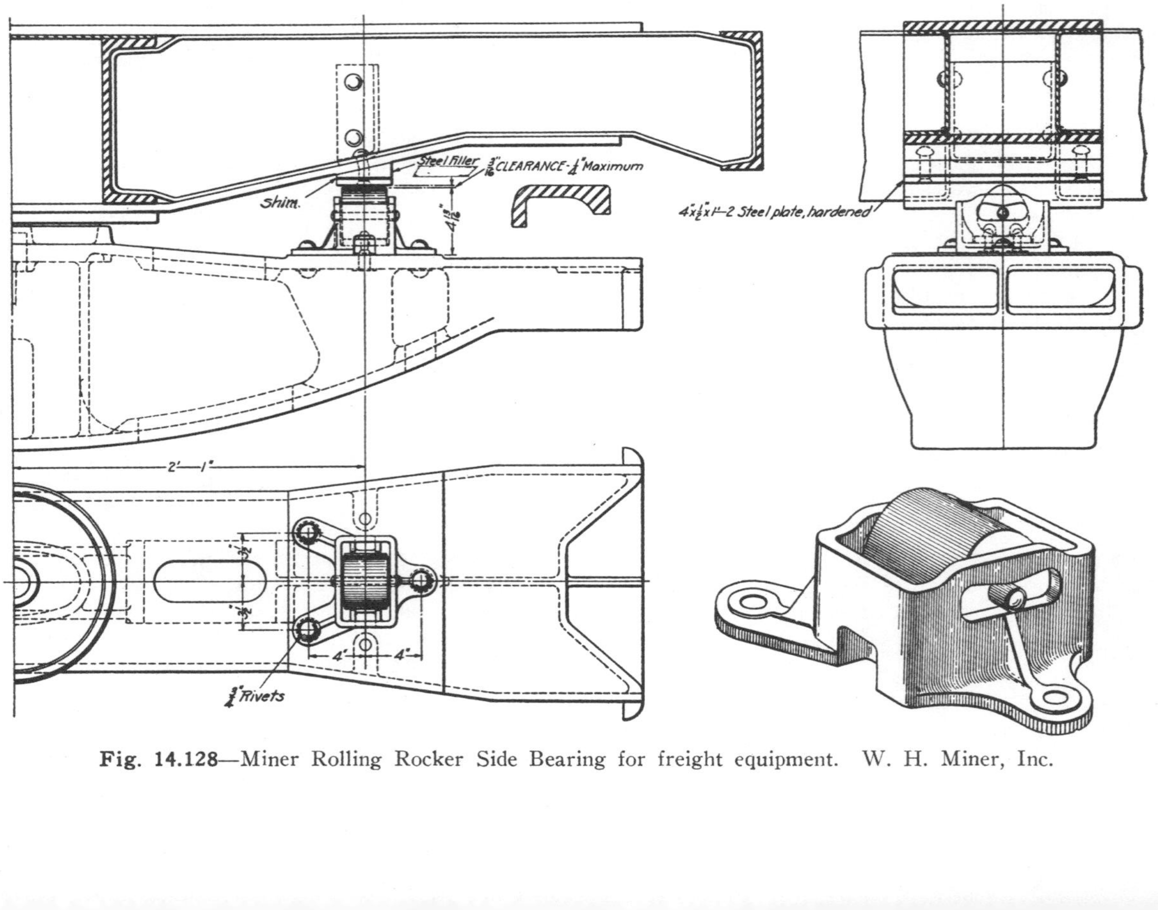

Fig. 14.128—Miner Rolling Rocker Side Bearing for freight equipment. W. H. Miner, Inc.

Woods Anti-Friction Side Bearings

For freight and passenger cars

ANTI-FRICTION side bearings manufactured by Edwin S. Woods and Company are designed for service under the severe conditions imposed by modern heavy, high-speed freight and passenger car equipment. Their function is to eliminate friction between the rubbing surfaces of car and truck. The existence of friction at these points retards free truck movement and increases train resistance. It causes excessive flange wear and contributes notably to the hazard of derailment with its undesirable consequences. Edwin S. Woods and Company have pioneered in the development of anti-friction side bearings. Their products have served with distinction practically every important railroad, since 1898.

Fig. 14.129—Woods Forged Steel Roller Side Bearing

Woods Forged Steel Side Bearing is constructed of only two parts: the housing and the roller. The housing is of special high quality S. A. E. forging steel, heat treated and floor quenched to insure a high Brinnell. The roller is forged from S. A. E. steel, heat treated and oil quenched, and is centered by gravity. There is minimum clearance between the housing side walls and the roller, assuring a true position of the roller at all times. The roller stops at the limit of travel are of the same contour as the roller so that upon reaching the limit of travel the roller contacts these stops which are one-quarter of its circumference. Woods Forged Steel Side Bearing is exceptionally rugged and is ideally suited to all types of rolling stock.

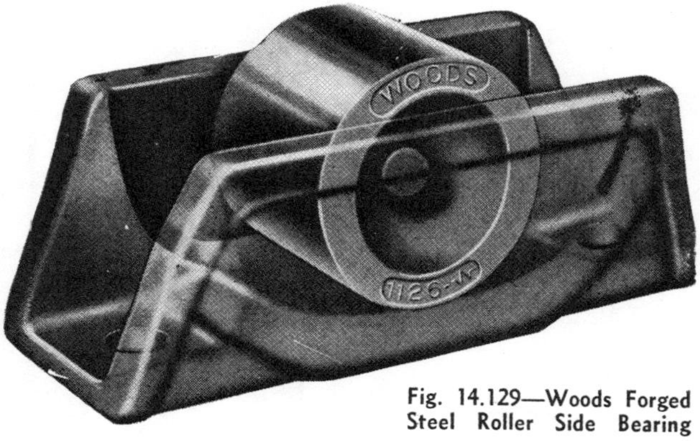

Fig. 14.129—Woods Forged Steel Roller Side Bearing

Woods Tip Roller Side Bearing for Application to Body Bolster. Fig. 14.130

The Woods "Tip Roller" Side Bearing is designed for application to the body bolster of freight cars and locomotive tenders. By the combination of a number of large intermeshing rollers, this side bearing secures any required carrying capacity, and at the same time allows sufficient travel for ordinary working conditions. Being suspended from the body bolster, the bearing is at all times free from foreign substances and the rollers center by gravity upon release of load. The casing is of certified malleable iron. Rollers of white iron operate between oil-tempered spring-steel plates.

Woods 7-in. Travel Gravity Roller Side Bearings. Fig. 14.124

This bearing consists of a forged steel, heat treated and oil quenched roller operating between a forged steel heat treated filler below and a spring steel bolster plate above. The roller is retained in a malleable iron housing by its trunnions and is centered by gravity. The trunnions extend into, but are free from engagement with, the side walls. The side walls of the housings are so designed that heavy sections of metal take care of the end thrust. This is a heavy duty bearing and provides for standard A. A. R. three rivet spacing.

Woods Counter-Balance Truck Bolster Side Bearing For Freight Cars. Fig. 14.126

Many railroads require a truck side bearing in order to maintain a standard height from the rail to the top of side bearing. In order to meet these conditions, the Counter-Balance Bearing was designed. This bearing is of sufficient capacity for the heaviest equipment. The casing is made of malleable iron; the rollers of forged steel, the lower half of the roller being increased in width sufficiently to immediately center it when the load is released. The wearing plate is of oil-tempered spring steel. This side bearing is similar to and interchangeable with the U.S.R.A. standard truck bolster side bearings.

Woods Spring Controlled Truck Side Bearing for Freight and Passenger Cars. Fig. 14.125

This side bearing has sufficient carrying capacity for the heaviest freight and passenger equipment. It has a travel of ten inches, which is sufficient for a 75-ft. car on a 26-deg. curve. The casing is of malleable iron, enclosing a carriage supplied with six or eight steel rollers, 2 in. in diameter by 1¼ in. long, operating between spring-steel wearing plates. The carriage is controlled by a spring which instantly centers it on release of load. This bearing is attached to the truck bolster and is always free from foreign substances which might interfere with its operation.

Woods Passenger Car Side Bearing, Showing Roller and Springs. Fig. 14.127

This bearing is generally known as the "Pullman" type passenger bearing, having been standard on all Pullman cars for a number of years. It is a single-roller, body side bearing with travel of ten inches, and is used on six-wheel truck passenger equipment. Casing is of malleable iron, roller of white iron, and wearing plate of oil-tempered spring steel.

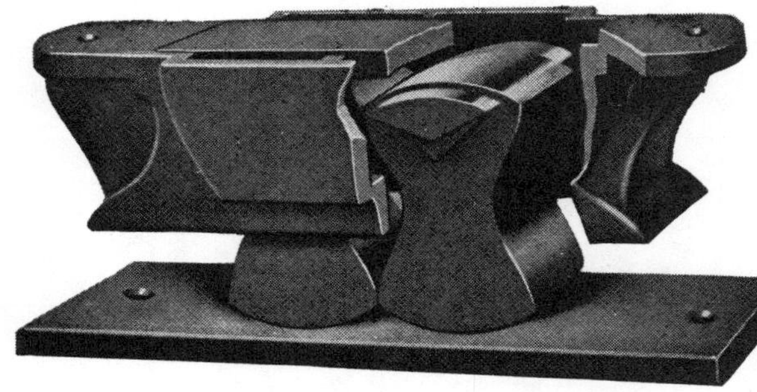

Fig. 14.130—Woods Tip Roller Side Bearing for application to body bolsters

EDWIN S. WOODS & COMPANY, 4710 W. Division St., CHICAGO, ILL.

A DIVISION OF PETTIBONE MULLIKEN CORPORATION

Other Products and Branch Offices Are Listed in the Classified Indexes

Fig. 14.131—Barber Monocast Adjustable Side Bearing.
(Housing cast integral with bolster)

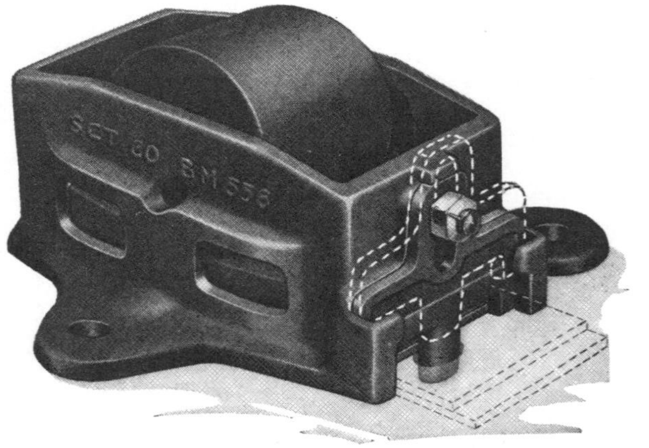

Fig. 14.132—Barber No. 17 Adjustable Roller Side Bearing.
(Riveted Type)

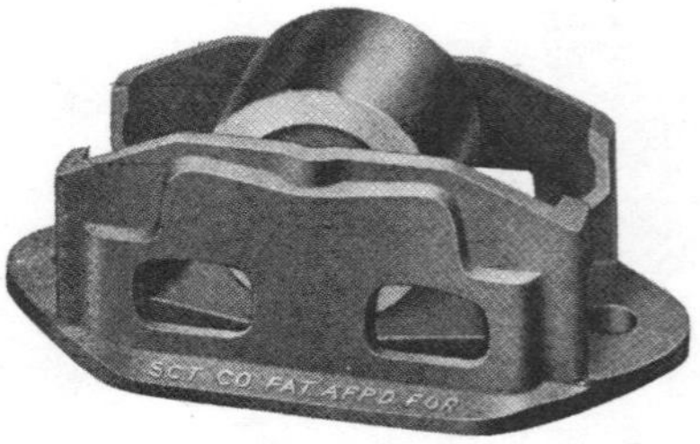

Fig. 14.133—Barber No. 21 Roller Side Bearing.
(Riveted Type)

Standard Car Truck Company

(*See also Pages* 1097, 1099)

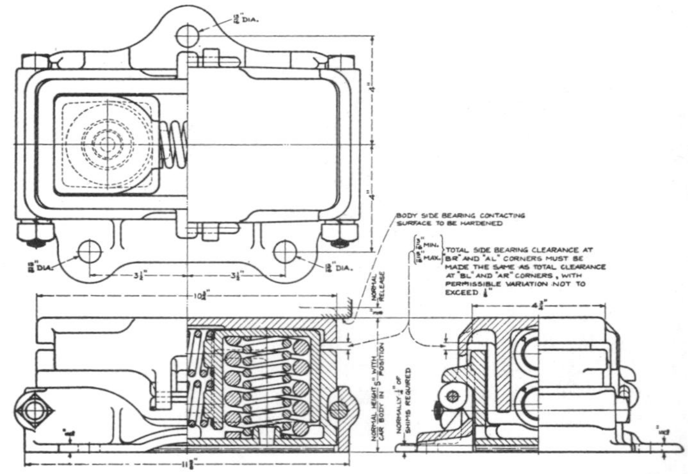

Fig. 14.134—Symington resilient constant contact non-harmonic side bearing for freight car trucks.

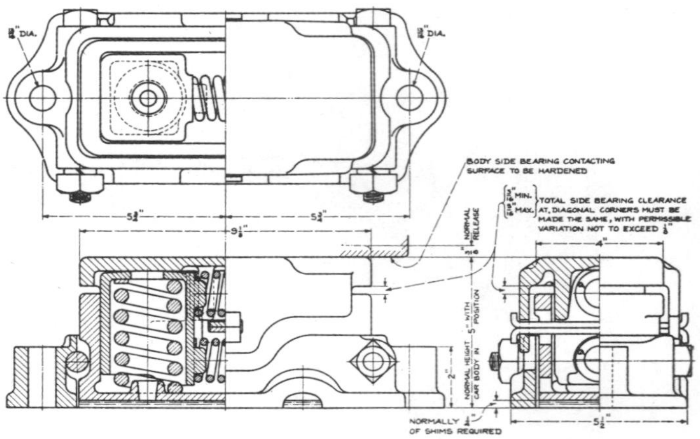

Fig. 14.135—Symington resilient constant contact non-harmonic side bearing for four-wheel passenger car trucks.

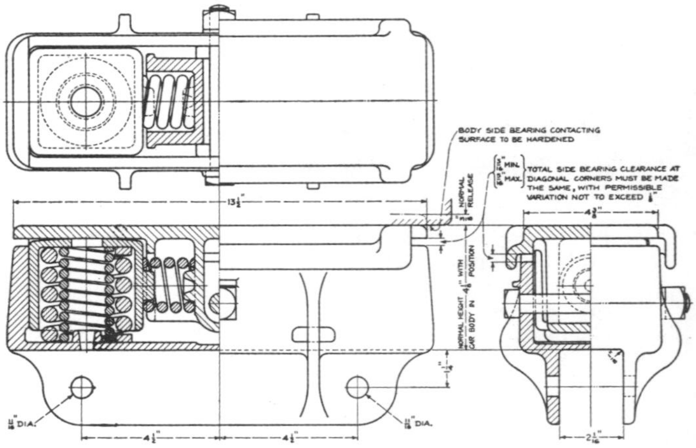

Fig. 14.136—Symington resilient constant contact non-harmonic side bearing for six-wheel passenger car trucks.

The Symington-Gould Corporation

(*See also Page* 1101)

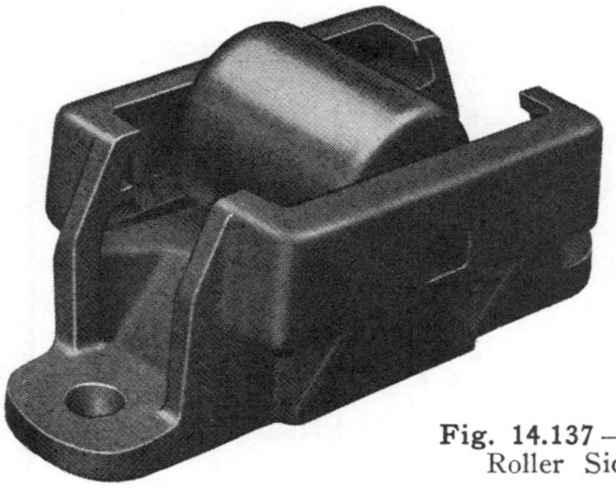

Fig. 14.137—Wine Single-Roller Side Bearing.

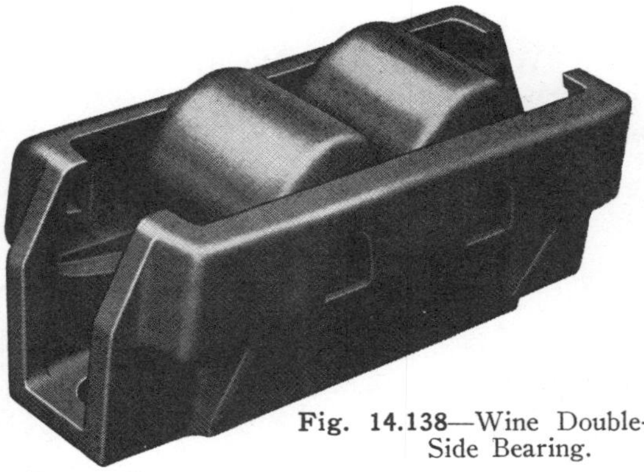

Fig. 14.138—Wine Double-Roller Side Bearing.

The Wine Railway Appliance Company

(*See also Page* 291)

Fig. 14.139—Symington Ball-Bearing Center Plates. The Symington-Gould Corporation.

Fig. 14.140—Symington Ball-Bearing Equalizing Center Plates. The Symington-Gould Corporation.

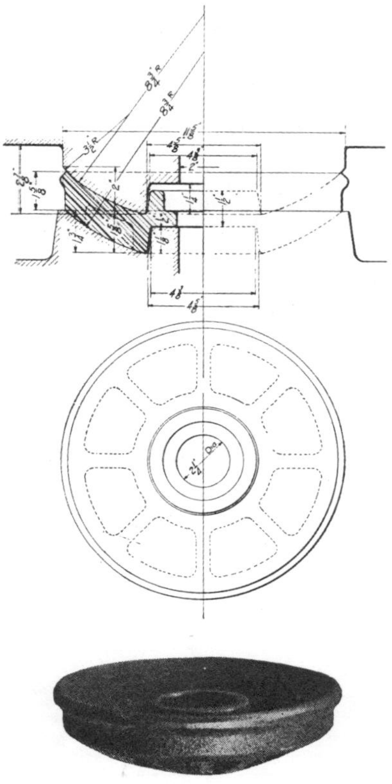

Fig. 14.141—Miner Center Plate Shim as Applied to Passenger Cars. W. H. Miner, Inc.

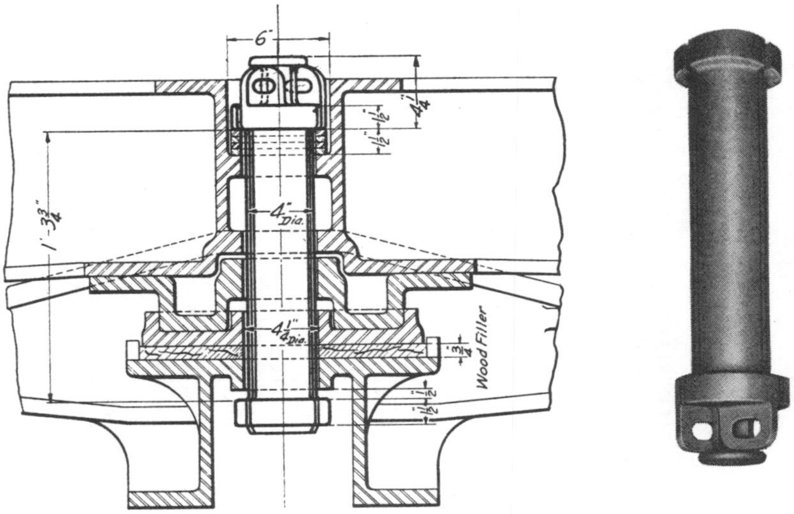

Fig. 14.142—Safety Bolster Locking Center Pin for Passenger Equipment. W. H. Miner, Inc.

Wheel and Axle Manual

The "Wheel and Axle Manual" (Revised 1935) which has been compiled under the auspices of the A.A.R. Committee on Wheels is a very complete compendium of A.A.R. specifications, standards and recommended practices. It includes information as to wheel manufacture, shop procedure, tests, inspections, defects, and use of A.A.R. gages.

A list of A.A.R. Specifications will be found in Section 19. Wheel tread and flange sections are shown below and sections of standard wheels on immediately following pages.

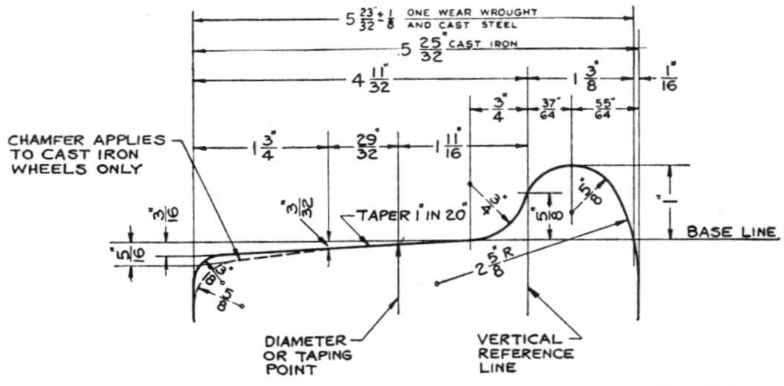

Fig. 14.143—Wheel Tread and Flange for Cast Iron, One-Wear Wrought and Cast Steel Wheels.
(Manual—Page D-40)

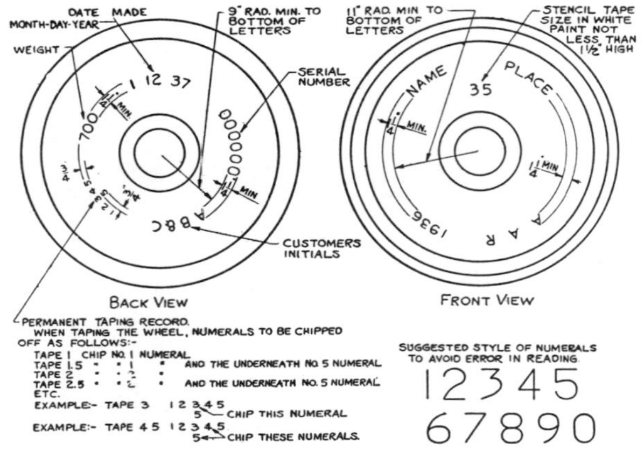

Fig. 14.144—Recommended Marking of Cast Iron Wheels (1937).

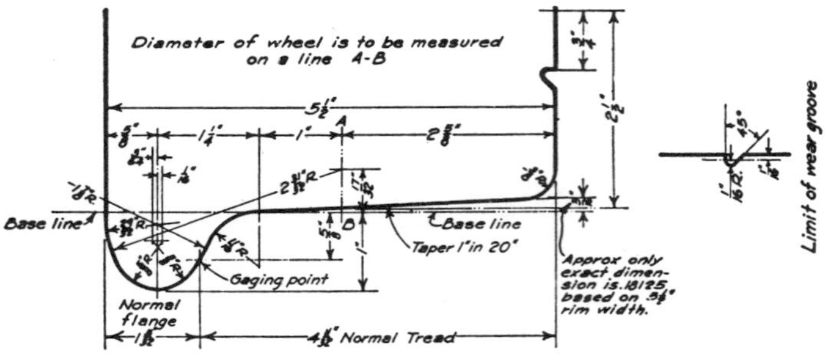

Fig. 14.145—A.A.R. Standard Wheel Tread and Flange for All Steel-Tired Wheels.
(Manual—Page D-50)

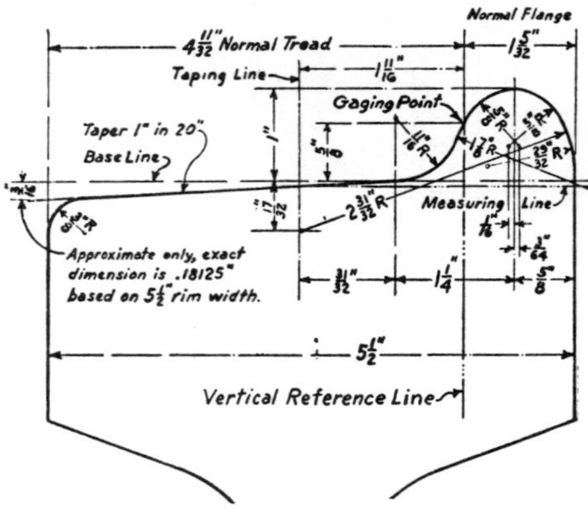

Fig. 14.146—Tread and Flange for Multiple and Two-Wear Wrought Steel Wheels.
(Manual—Page D-50A)

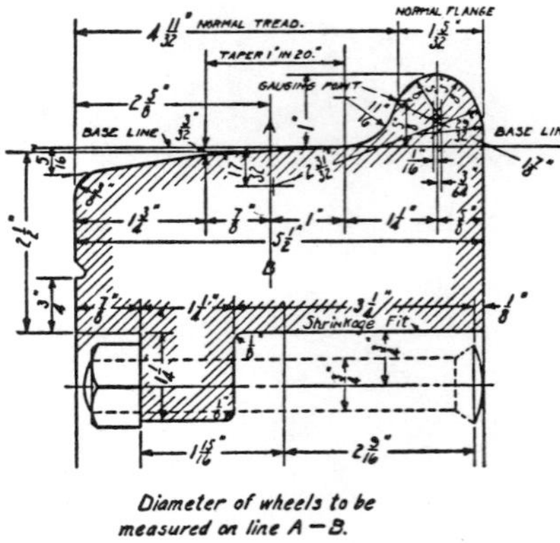

Fig. 14.147—A.A.R. Standard Tire Fastening for Steel Tired Wheels.
(Manual—Page D-51)

A.A.R. New Passenger Car Axles.
(See Fig. 14.151, Page 1136)

Tubular Railway Axles for All Purposes.
(Manual—Page D-6A, Adopted 1941)
(See also Pages 1142, 1143)

Classification of Axle	Size of Journal	Dimensions																	
		A	B	C	D	E	F	G	H	I	J	K	L	M	N	O	P	Q	R
A	3 3/4" × 7"	5/8"	7"	2 1/2"	8 1/16"	13/16"	23/32"	1 1/2"	4 3/4"	3 3/4"	4 3/4"	5 1/8"	5 1/2"	4 7/8"	4 1/4"	3'-9 1/4"	5'-3"	6'-3"	6'-11 1/4"
B	4 1/4" × 8"	5/8"	8"	2"	8 1/16"	13/16"	23/32"	1 1/2"	5 1/4"	4 1/4"	5 1/4"	5 3/4"	6 1/8"	5 7/16"	4 3/4"	3'-9 1/4"	5'-3"	6'-3"	7'-0 1/4"
C	5" × 9"	3/4"	9"	2"	8 1/16"	13/16"	23/32"	1 1/2"	6 1/8"	5"	6 1/8"	6 1/2"	6 7/8"	6 3/16"	5 3/8"	3'-9 1/4"	5'-3"	6'-4"	7'-2 1/2"
D	5 1/2" × 10"	3/4"	10"	2"	8 1/16"	13/16"	23/32"	3"	6 5/8"	5 1/2"	6 5/8"	7"	7 3/8"	6 3/4"	5 7/8"	3'-9 1/4"	5'-3"	6'-5"	7'-4 1/2"
E	6" × 11"	7/8"	11"	2 1/4"	8 5/8"	1 3/4"	3/4"	3"	7 1/4"	6"	7 1/4"	7 5/8"	8"	7 3/16"	6 7/16"	3'-5 3/4"	5'-2 1/2"	6'-6"	7'-6 3/4"
F	6 1/2" × 12"	7/8"	12"	2 1/4"	8 5/8"	1 3/4"	3/4"	3"	7 3/4"	6 1/2"	7 3/4"	8 1/8"	8 1/2"	7 13/16"	6 7/8"	3'-5 3/4"	5'-2 1/2"	6'-7"	7'-8 3/4"

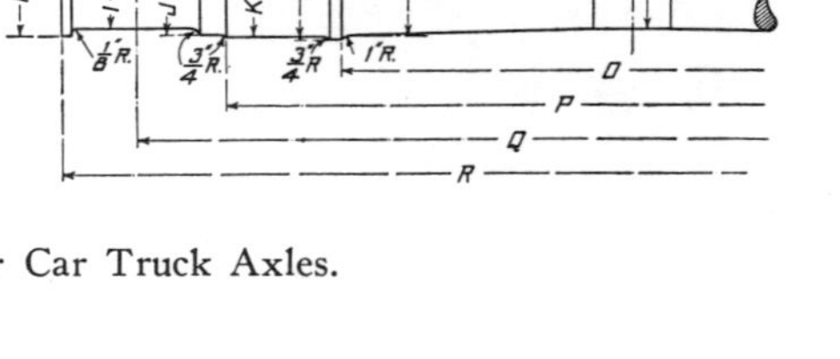

Fig. 14.148—A.A.R. Standard Dimensions for Car Truck Axles.
(Manual—Page D-6)
(See also Fig. 2884)

M-H Electronic Alarm System

Provides automatic indication of dangerous bearing temperatures
Train delays and injury to equipment avoided

THE critical nature of high-speed passenger train operation make it imperative that all practical measures be taken to head-off trouble before the danger point is reached. This is particularly true of the journal bearings of cars and locomotives. To afford protection for these important units, Minneapolis-Honeywell has developed an Electronic Bearing Alarm System. This System has been under road test in actual railroad operation for about a year and is being applied to additional equipment.

Apparatus and How It Functions

The System consists essentially of a temperature element applied to each bearing with a connection to a conveniently located control box. A motor-driven selector switch shown in one of the illustrations connects each couple progressively with the temperature measuring element in the control box. Should the temperature in any bearing exceed the predetermined value a lamp on the control box corresponding to that bearing will light and an alarm bell will sound. The light will remain lighted and the selector switch will stop. The light may be extinguished by manual operation of a reset switch, or by reducing the temperature of the bearing through the addition of oil, reduction of speed, or other means. If desired, the temperature of any bearing can be determined at any time by means of a manually-operated switch. Temperature measurement is accurate to within plus or minus 5 deg. F.

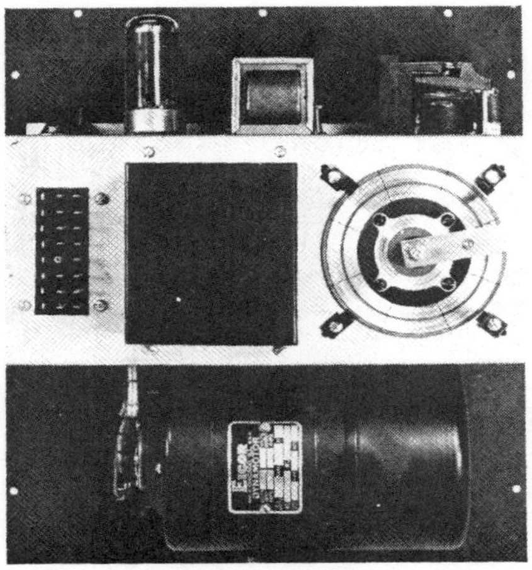

Interior of cabinet showing selector switch, motor-generator, and various relays

On the systems now in service the motor-driven selector switch is operated at a speed which will check the temperature on each individual bearing once each minute. The alarm point is set as low as possible consistent with running temperatures, usually about 180 deg., so that the alarm will ring soon enough on a temperature rise, to permit the train crew to take corrective action before a critical temperature is reached. For example, if the melting point of the babbitt in a friction type bearing is 350 degrees, a warning that the temperature of that bearing is rising, given at 180 deg., is usually soon enough to allow the train crew to take necessary measures to cool the journal bearing before the contact surfaces are damaged. The same procedure is applicable to anti-friction bearings. The presence of an open circuit or short circuit is indicated automatically and in such a case the temperature element in that particular journal can be cut out and protection retained on the others.

Source of Power

Power for operating this device is obtained from the battery circuit and the power consumption is small. A small self-contained motor generator set converts this d.c. supply to a.c. for the temperature measurement system. The cabinet which contains all the equipment except the temperature bulbs measures $12\frac{3}{8}$ by $13\frac{1}{4}$ by $8\frac{1}{4}$ in. The temperature bulbs applied to bearings are $3\frac{3}{8}$ in. long by $\frac{5}{8}$ in. in diameter and will fit into any standard journal casting. Other shapes and sizes of temperature elements designed for a number of different types of temperature measurements are also available.

Outside view of cabinet, showing signal lights corresponding to various bearings

General Uses

In addition to its primary use, the M-H Electronic Alarm System is also offered for measuring and signaling temperatures of cylinder walls, Diesel or steam, main engine bearings, traction motor bearings, cooling water, lubricating oil, and exhaust gases on locomotives.

MINNEAPOLIS-HONEYWELL REGULATOR COMPANY, MINNEAPOLIS, MINN.

Other Products and Branch Offices Are Listed in the Classified Indexes

Springs

The principal springs on railway freight and passenger cars are those used in truck construction and in draft gear; they may be of either elliptical or coil types. The following specifications covering their design and manufacture have been adopted by the Mechanical Division of the Association of American Railroads; all of which may be found in the Manual, Section A.

Elliptical Springs M-113
Helical Springs M-114
Carbon Steel Bars for Railway Springs . . M-112

Groups of coil springs for freight car trucks have been standardized. These are shown in the Manual and are illustrated herewith.

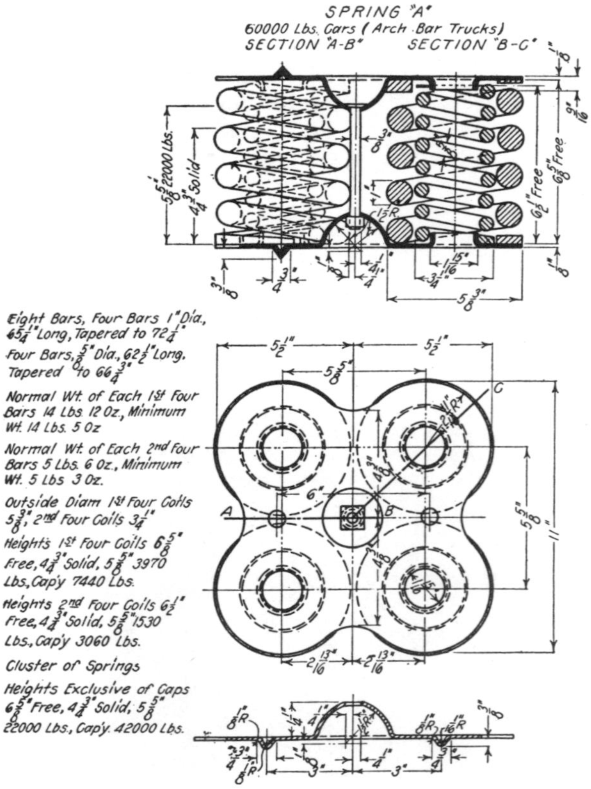

Fig. 14.153—A.A.R. Spring "A" and Caps.

(Manual—Page A-27)

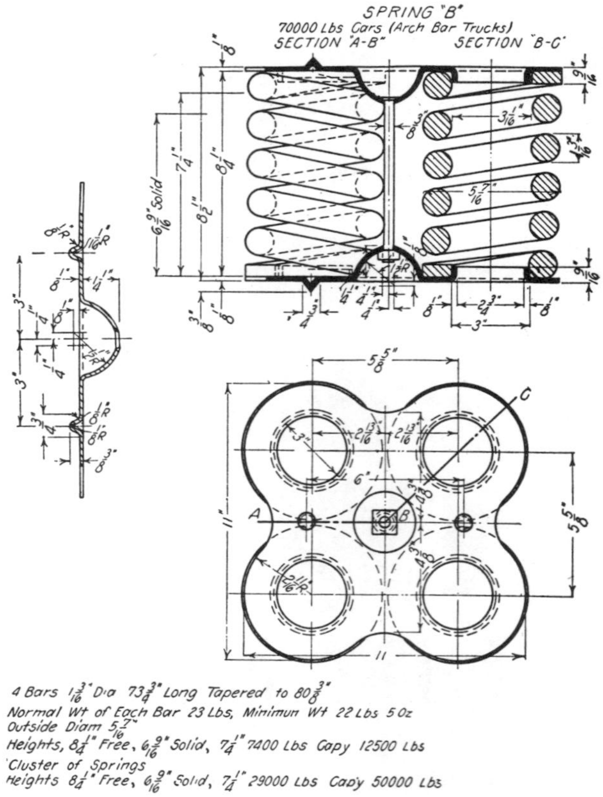

Fig. 14.154—A.A.R. Spring "B" and Caps.

(Manual—Page A-28)

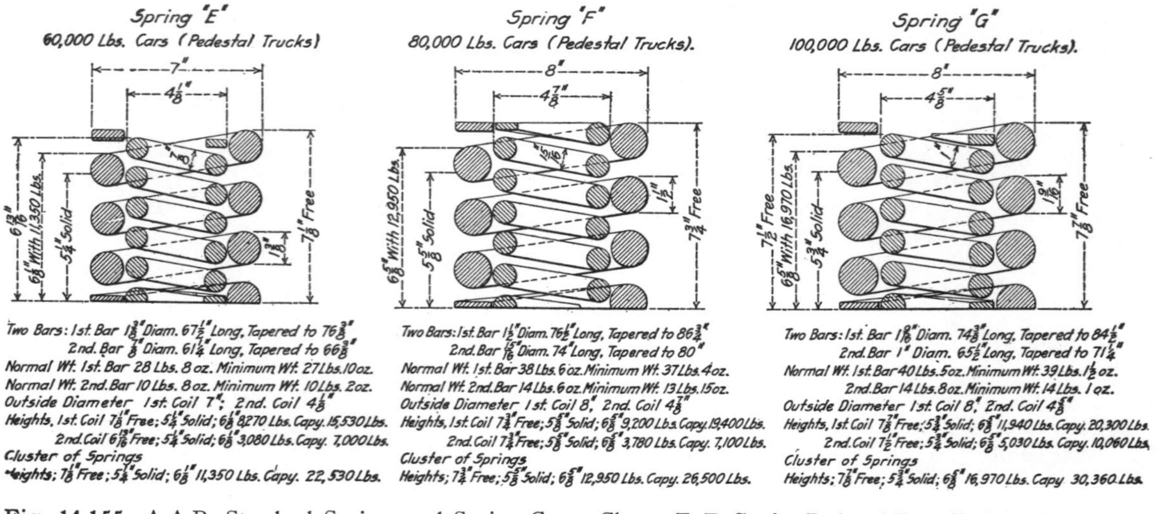

Fig. 14.155—A.A.R. Standard Springs and Spring Caps—Classes E, F, G—for Pedestal Type Freight Car Trucks.

(Manual—Pages D-31 and D-32)

(See also Page 1146)

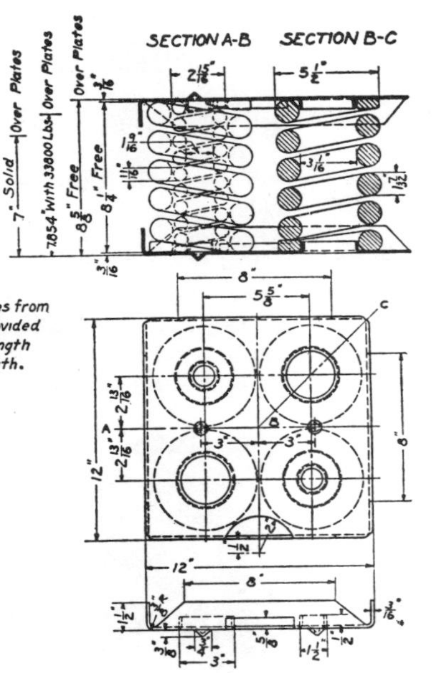

Note:
Other designs of spring plates from those shown may be used provided they are of equivalent strength and interchangeable therewith.

Six Bars. Four Bars $1\frac{7}{32}$" Dia. $73\frac{1}{8}$" Long. Tapered To $79\frac{5}{8}$".
Two Bars $\frac{11}{16}$" Dia. $68\frac{1}{8}$" Long. Tapered To $71\frac{5}{8}$".
Normal Wt. of Each 1st Four Bars 24 Lbs. 3 Oz. Minimum 23 Lbs. 8 Oz.
Normal Wt. of Each 2nd Two Bars 7 Lbs. 3 Oz. Minimum 7 Lbs. 0 Oz.
Outside Diam. of Outer Coils $5\frac{1}{2}$". Inner Coil $2\frac{15}{16}$".
Heights: Outer Coils $8\frac{1}{4}$" Free - $6\frac{5}{8}$" Solid - At 7.479" Load - 7229 Lbs. Capy. Solid - 15236 Lbs.
Heights: Inner Coils $8\frac{1}{4}$" Free - $6\frac{5}{8}$" Solid - At 7.479" Load - 2442 Lbs. Capy. Solid - 5147 Lbs.
Cluster of Springs - Total Normal Wt. Cluster Complete Including Plates 130 Lbs.
Cluster of Springs - At 7.479" Load - 33800 Lbs. Capy. Solid - 71238 Lbs.

Fig. 14.156—Spring "C-2"—Used with A.A.R.-C (5 in. by 9 in.) Axle.
(Manual—Page D-29)

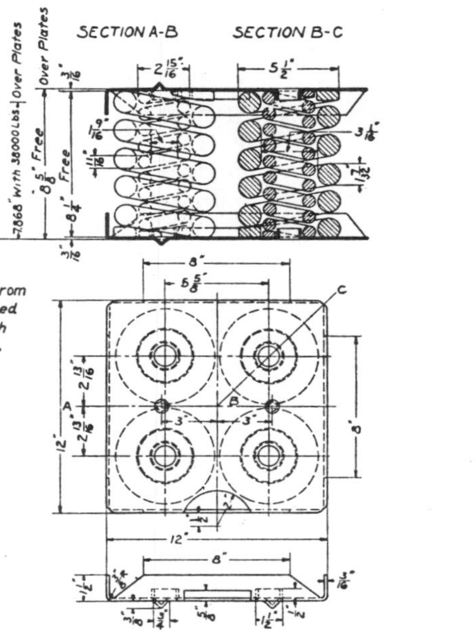

Note:
Other designs of spring plates from those shown may be used provided they are of equivalent strength and interchangeable therewith.

Eight Bars. Four Bars $1\frac{7}{32}$" Dia. $73\frac{1}{8}$" Long. Tapered To $79\frac{5}{8}$".
Four Bars $\frac{11}{16}$" Dia. $68\frac{1}{8}$" Long. Tapered To $71\frac{5}{8}$".
Normal Wt. of Each 1st Four Bars 24 Lbs. 3 Oz. Minimum 23 Lbs. 8 Oz.
Normal Wt. of Each 2nd Four Bars 7 Lbs. 3 Oz. Minimum 7 Lbs. 0 Oz.
Outside Diam. of Outer Coils $5\frac{1}{2}$". Inner Coil $2\frac{15}{16}$".
Heights: Outer Coils $8\frac{1}{4}$" Free - $6\frac{5}{8}$" Solid - At 7.493" Load - 7100 Lbs. Capy. Solid - 15236 Lbs.
Heights: Inner Coils $8\frac{1}{4}$" Free - $6\frac{5}{8}$" Solid - At 7.493" Load - 2400 Lbs. Capy. Solid - 5147 Lbs.
Cluster of Springs - Total Normal Wt. Cluster Complete Including Plates 144 Lbs.
Cluster of Springs - At 7.493" Load - 38000 Lbs. Capy. Solid - 81532 Lbs.

Fig. 14.157—Spring "D-2"—Used with A.A.R.-D ($5\frac{1}{2}$ in. by 10 in.) Axle.
(Manual—Page D-30)

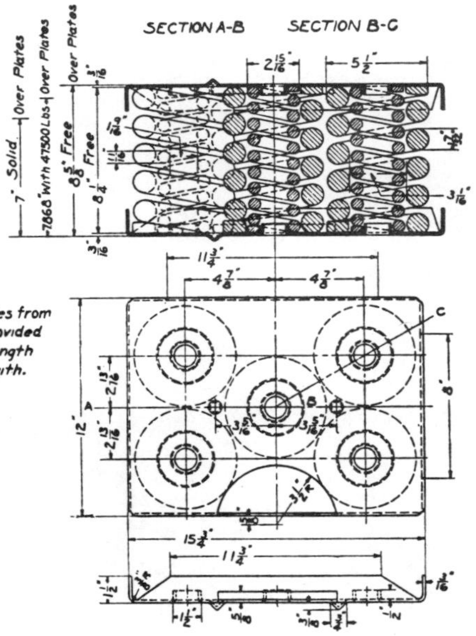

Note:
Other designs of spring plates from those shown may be used provided they are of equivalent strength and interchangeable therewith.

Ten Bars. Five Bars $1\frac{7}{32}$" Dia. $73\frac{1}{8}$" Long. Tapered To $79\frac{5}{8}$".
Five Bars $\frac{11}{16}$" Dia. $68\frac{1}{8}$" Long. Tapered To $71\frac{5}{8}$".
Normal Wt. of Each 1st Five Bars 24 Lbs. 3 Oz. Minimum 23 Lbs. 8 Oz.
Normal Wt. of Each 2nd Five Bars 7 Lbs. 3 Oz. Minimum 7 Lbs. 0 Oz
Outside Diam. of Outer Coils $5\frac{1}{2}$". Inner Coil $2\frac{15}{16}$".
Heights: Outer Coils $8\frac{1}{4}$" Free - $6\frac{5}{8}$" Solid - At 7.493" Load - 7100 Lbs. Capy. Solid - 15236 Lbs.
Heights: Inner Coils $8\frac{1}{4}$" Free - $6\frac{5}{8}$" Solid - At 7.493" Load - 2400 Lbs. Capy. Solid - 5147 Lbs.
Cluster of Springs - Total Normal Wt. Cluster Complete Including Plates 181 Lbs.
Cluster of Springs - At 7.493" Load - 47500 Lbs. Capy. Solid - 101915 Lbs.

Fig. 14.158—Spring "H-2"—Used with A.A.R.-E (6 in. by 11 in.) Axle.
(Manual—Page D-33)

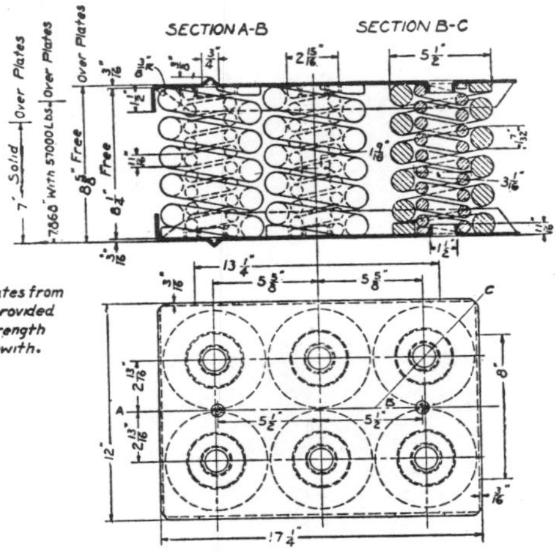

Note:
Other designs of spring plates from those shown may be used provided they are of equivalent strength and interchangeable therewith.

Twelve Bars. Six Bars $1\frac{7}{32}$" Dia. $73\frac{1}{8}$" Long. Tapered To $79\frac{5}{8}$".
Six Bars $\frac{11}{16}$" Dia. $68\frac{1}{8}$" Long. Tapered To $71\frac{5}{8}$".
Normal Wt. of Each 1st Six Bars 24 Lbs. 3 Oz. Minimum 23 Lbs. 8 Oz.
Normal Wt. of Each 2nd Six Bars 7 Lbs. 3 Oz. Minimum 7 Lbs. 0 Oz.
Outside Diam. of Outer Coils $5\frac{1}{2}$". Inner Coil $2\frac{15}{16}$".
Heights: Outer Coils $8\frac{1}{4}$" Free - $6\frac{5}{8}$" Solid - At 7.493" Load - 7100 Lbs. Capy. Solid - 15236 Lbs.
Heights: Inner Coils $8\frac{1}{4}$" Free - $6\frac{5}{8}$" Solid - At 7.493" Load - 2400 Lbs. Capy. Solid - 5147 Lbs.
Cluster of Springs - Total Normal Wt. Cluster Complete Including Plates 217 Lbs.
Cluster of Springs - At 7.493" Load - 57000 Lbs. Capy. Solid - 122298 Lbs.

Fig. 14.159—Spring "P-2"—Used with A.A.R.-F ($6\frac{1}{2}$ in. by 12 in.) Axle.
(Manual—Page D-34)

(See also Page 1145)
(For Snubbers see Pages 1151-1162)

Springs of All Types for All Classes of Equipment

Facilities of American Steel Foundries assure dependability

SPRINGS manufactured for railway service by American Steel Foundries include all types,—helical, volute, and elliptic, for passenger and treight car trucks, caboose car trucks, draft gears and all miscellaneous purposes.

Plant Facilities

The American Steel Foundries spring plant at Hammond, Indiana, is equipped with up-to-date machinery and pyrometer-controlled heat treating furnaces, assuring the initial heating of spring steel and the drawing operation of the springs at proper temperatures.

These facilities, together with the long and intimate contact with spring problems incident to the operation of modern railway rolling stock, enable the Company to furnish springs of distinctly superior quality, and which will render long and dependable service.

Spring clips and Potter End Castings are furnished in cast steel. Bands are furnished in forged and cast steel. When required, springs are manufactured in alloy steels.

Technical Staff

The technical staff of American Steel Foundries is always available to railway mechanical staffs for consultation in connection with any problem relating to spring designs for any class of rolling equipment.

Examples of heavy-duty volute and helical springs manufactured by American Steel Foundries.

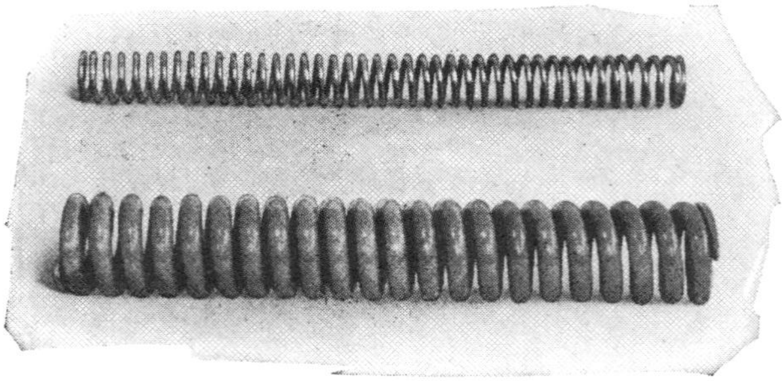

Single-coil compression springs.

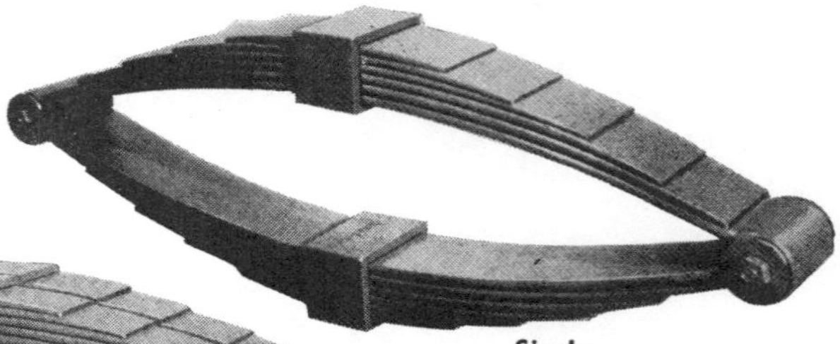

Single full-elliptic spring.

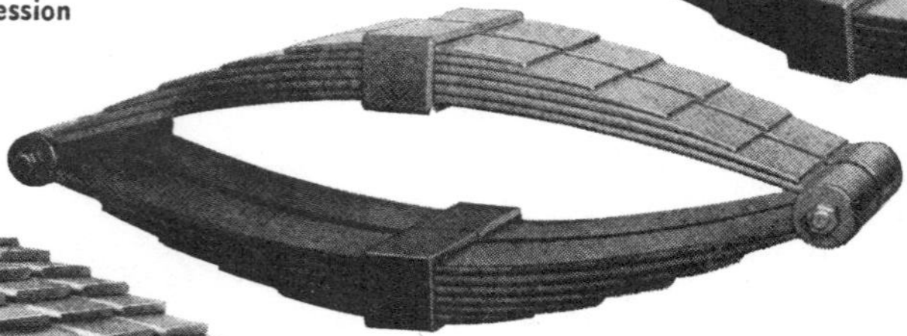

Duplicate full-elliptic passenger car and tender truck spring.

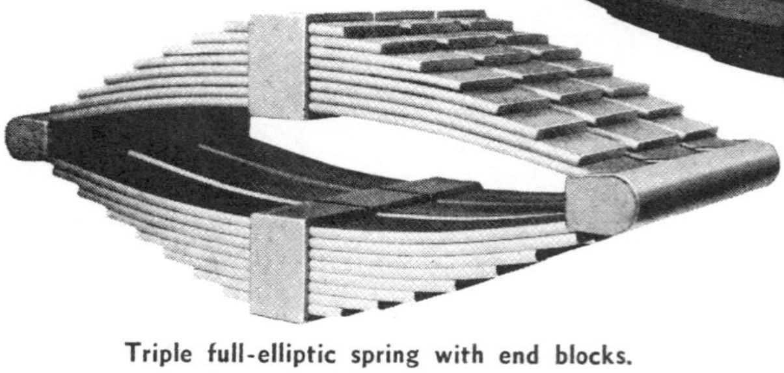

Triple full-elliptic spring with end blocks.

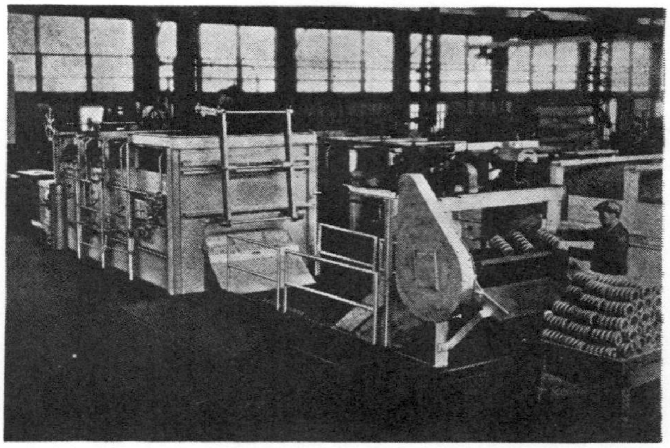

Above—Heat-treating furnace for helical springs. To the left—Tempering furnace for drawing helical or elliptic springs.

AMERICAN STEEL FOUNDRIES, CHICAGO, ILL.

Other Products and Branch Offices Are Listed in the Classified Indexes

Springs of All Types for Freight and Passenger Cars

THE Railway Steel-Spring Division of the American Locomotive Company manufactures all types of springs for use in foreign and domestic car equipment. The Company has Spring Plants advantageously situated and equipment with the most modern and up-to-date appliances, together with special machinery and modern automatically regulated furnaces for the manufacture and heat-treating of springs of the highest grade and serviceability. The company is also fully equipped to furnish springs made of various alloy steels.

Quintuple Elliptic Spring
Suitable for four-wheel trucks

The easy and comfortable riding of cars in passenger train service depends largely upon the exact suitability of the springs for the loads involved. The experienced engineers of the American Locomotive Company have been eminently successful in designing springs for the exacting and unusual requirements of modern passenger train service. Many of the modern high-speed trains owe their fine-riding success to the smooth functioning of springs designed and manufactured by the American Locomotive Company.

Equalizer Springs

Equalizing Springs are of as much importance as the Elliptical Springs and must receive just as great care in design and manufacture. They are subject to the first effects of all jolts, and oscillations not absorbed are passed through to the other springs and the car body.

For trucks of designs other than the Standard Dropped Equalizer Bar type, the spring designs often present problems requiring most skillful engineering. This company is prepared to offer expert engineering service on all problems involving spring design and manufacture.

2-Coil
Draft Gear Spring

Freight Car Springs

The Standard A.A.R. Helical Springs are used on the majority of present-day cars. For the conventional type of truck the standard groups are suitable. For the Double-Truss type of truck the Standard Helical Springs are used and grouped to suit the available space and car capacity.

Double Elliptic Spring
Suitable for six-wheel trucks

3-Coil
Equalizer Spring

5-Coil Group
with Universal UX-5 Plates

Draft Gear Springs

Draft Gear Springs have an important function to perform and should be made by the best possible process.

AMERICAN LOCOMOTIVE COMPANY, Railway Steel-Spring Div., NEW YORK, N. Y.

Other Products and Branch Offices Are Listed in the Classified Indexes

Universal Bolster Spring Plates

"Universal" in fact as in name

EACH of the designs of Universal Bolster Spring Plates is suited to receive any of the current combinations of A.A.R. Coil Springs of the 1915 or the 1936 designs, with or without friction springs or snubbers.

UC-4 Universal Spring Plate

For use on 30-, 35-, 40-, and 50-ton cars with the conventional A.A.R. type trucks, the UC-4 Universal Plate applies. Only the one kind of plate need be purchased and stocked for application to all of these cars.

These plates are equally "Universal" for all-coil groups or for groups with friction springs.

Other designs are available for the five-coil and six-coil groups used under cars of 70-ton capacity with Conventional Type Trucks.

UX-5 Universal Spring Plates for Double-Truss Trucks

Only one design of Universal Spring Plate is required for 35-, 40-, 50-, and 55-ton cars using the Double-Truss Trucks with or without spring planks. This plate is suitable with or without snubbers.

Additional designs are equally suitable for cars of 70-ton capacity or over.

All Universal Spring plates are made of heavy wrought steel amply flanged, and formed in precision presses, to retain the springs securely in place under all conditions and assure proper fitting in the trucks. They are particularly rugged and will withstand the service conditions and attendant corrosion for a long period. For specially corrosive service, plates can be made of a corrosion resistant steel.

UC-4 Universal Spring Plate for 4-Coil Groups

4-Coil Spring Group assembled in UC-4 Universal Spring Plates

UX-5 Universal Spring Plate for 5-Coil Groups

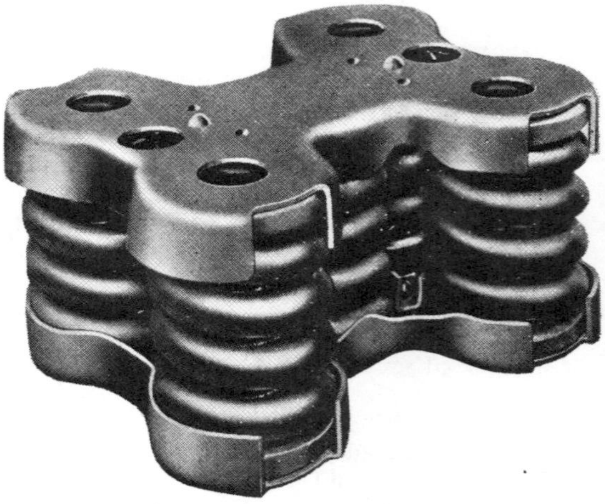

5-Coil Spring Group assembled in UX-5 Universal Spring Plate

The following companies are licensed to manufacture Universal Spring Plates:

AMERICAN LOCOMOTIVE CO., Railway Steel Spring Div., New York, N. Y.

AMERICAN SPIRAL SPRING & MFG. CO. Pittsburgh, Pa.
AMERICAN STEEL FOUNDRIES Chicago, Ill.
CRUCIBLE STEEL COMPANY OF AMERICA, New York, N. Y.
FORT PITT SPRING CO. Pittsburgh, Pa.
PITTSBURGH SPRING & STEEL CO. Pittsburgh, Pa.
UNION SPRING & MFG. CO. Pittsburgh, Pa.

STANDARD STEEL WORKS DIVISION—BALDWIN LOCOMOTIVE WORKS, Philadelphia, Pa.

Union Car Truck Equipment

Springs, Steel Castings, Pedestal Liners and Wear Plates
Journal Box Lids—All classes of Spring Plates

Group springs—all types available with Universal Spring Plates

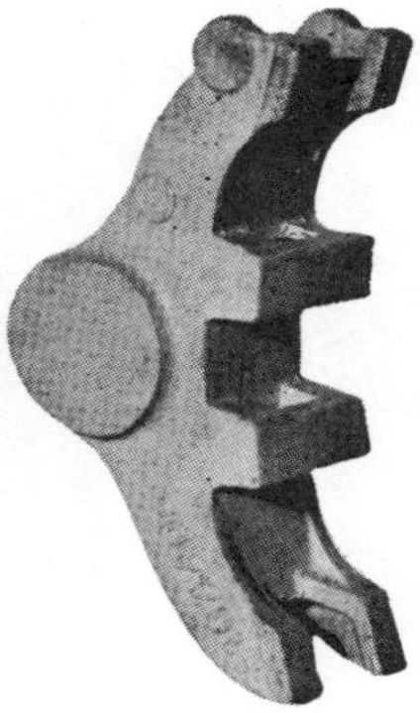

Cast Steel Brake Head

Truck springs built to specifications

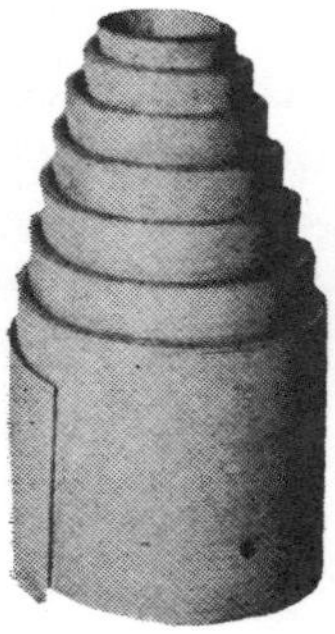

Volute Springs manufactured in sizes up to 9-in. by $\frac{7}{8}$-in. material

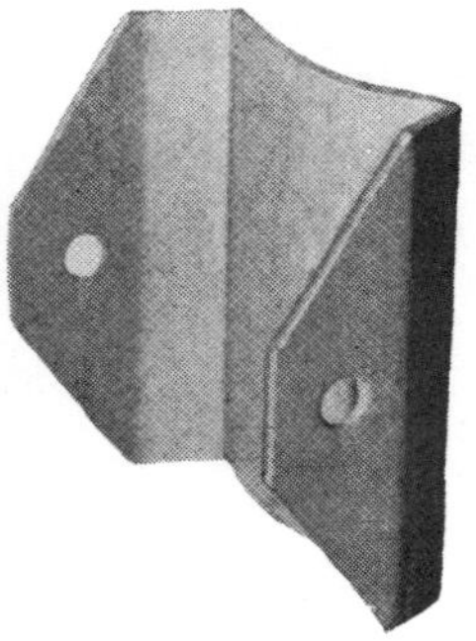

Left:—Heat treated steel Pedestal Liner

To the right:—Helical Spring

THE high speeds at which passenger and freight trains now operate have greatly increased the severity of service conditions which truck springs and other truck parts must meet. Recognizing the necessity for correspondingly higher standards, Union Spring and Manufacturing Company have developed their plant equipment to include the latest modern machinery including Magnaflux testing apparatus and a special shot blasting machine for Helical Springs.

Union Springs for Modern Railroading

One department of this manufacturer's plant is fully equipped for scientific heat-treating and is devoted especially to the manufacture of alloy steel springs. The Coil Department and the Elliptic Department have all the equipment necessary for supplying any railroad promptly with any type of spring.

Universal Spring Plates

Several designs of Universal Spring Plates are available—for use with 4-coil spring groups and for the 5-, 6-, and 7-coil groups of double truss trucks, also any combination of the various friction or shock-absorbing springs.

Journal Box Lids

The A. A. R. journal box lid illustrated herewith, is ruggedly constructed of $\frac{5}{32}$-in. steel and is 100 per cent dustproof. It is equipped with this Company's special spring and working parts.

This manufacturer is fully equipped to meet the most exacting railroad specifications in supplying all types of the products represented in the accompanying illustrations.

Wear Plate for side bearings and truck frames

A. A. R. Journal Box Lid

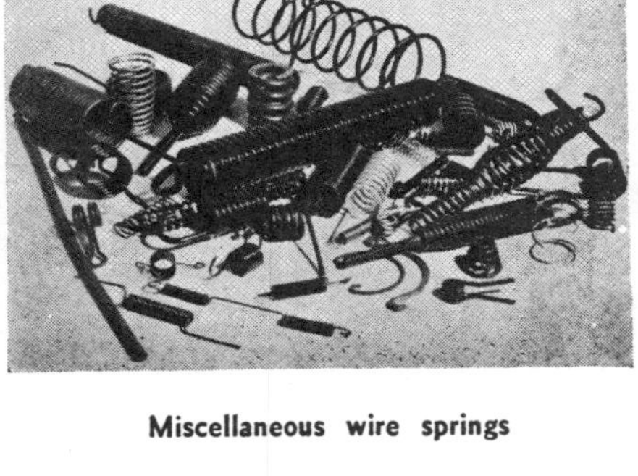

Miscellaneous wire springs

UNION SPRING AND MANUFACTURING CO., NEW KENSINGTON, PA.

Other Products and Branch Offices Are Listed in the Classified Indexes

Simplex Unit Type Snubber

For Freight Equipment

SERVICE has proved that the Simplex Unit Type Snubber makes easy riding freight cars.

It is designed to operate in combination with standard coil springs.

It controls coil-spring action by dampening or snubbing the harmonic oscillations that coils would otherwise develop at critical speeds and imparts a soft, smooth action to spring groups at all speeds.

It decreases lading damage claims and reduces the cost of car and roadbed maintenance.

The Simplex Unit Type Snubber, as illustrated, is interchangeable with any A. A. R. Class "D" coil spring in any freight car spring arrangement using any type of spring plate.

Application of Simplex Unit Type Snubber to A. A. R. Conventional Truck spring group

Principles of Design

The basic principle of Simplex Unit Type Snubber design is four simple friction wedges held together by uniformly distributed pressures created by the use of a rubber spring.

The Simplex Unit Type Snubber consists of five parts:—two interchangeable heat treated, alloy steel follower wedges; two interchangeable special grade cast iron side wedges; and one rubber spring—all interlocked together without the use of bolts. Heat treated steel working against cast iron provides an ideal combination for dissipating shock and resisting wear.

Quickly Inspected

The friction wedges of the Simplex Snubber are readily visible and can be inspected in place while under the car without loss of time.

The Simplex Unit Type Snubber has the following advantages:—

Simplex Design—Few parts (only five)—Long life — Easily inspected — Proved riding qualities — Replaces any spring unit—Rugged construction—Assembly interlocks without bolts.

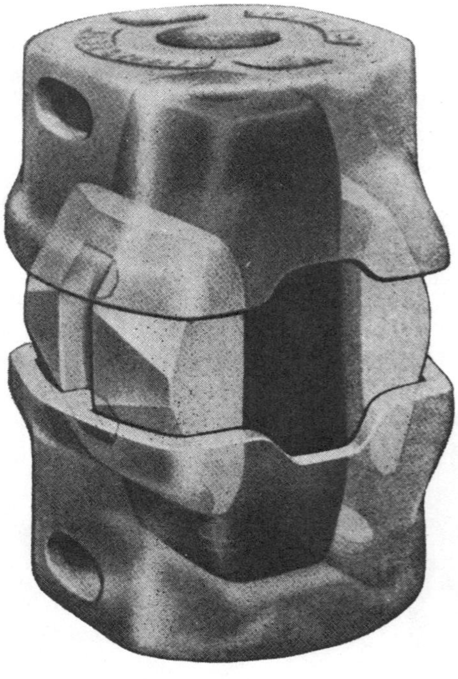

Phantom view of Simplex Unit Type Snubber

Five Parts of Simplex Unit Type Snubber

Fig. 14.160—Detail Parts of Cardwell Friction Bolster Spring, Type A.

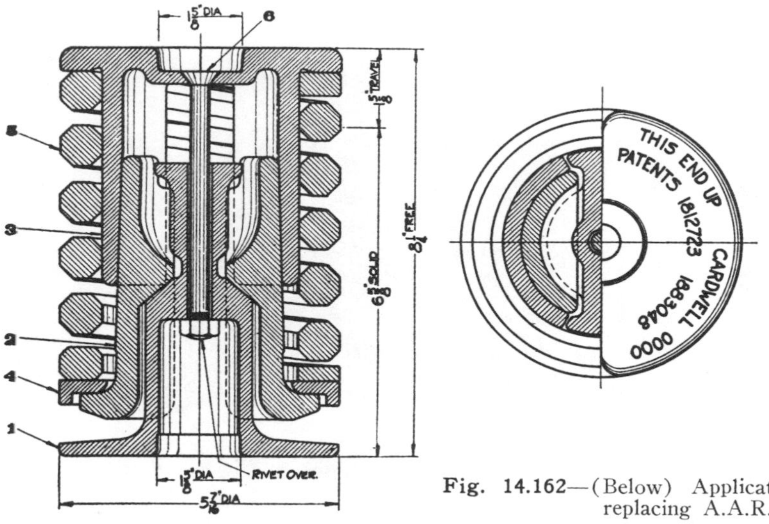

Fig. 14.161—(Left) Sectional drawing of Cardwell Friction Bolster Spring, Type A.

Fig. 14.162—(Below) Application of Cardwell Bolster Spring replacing A.A.R. Type D Spring.

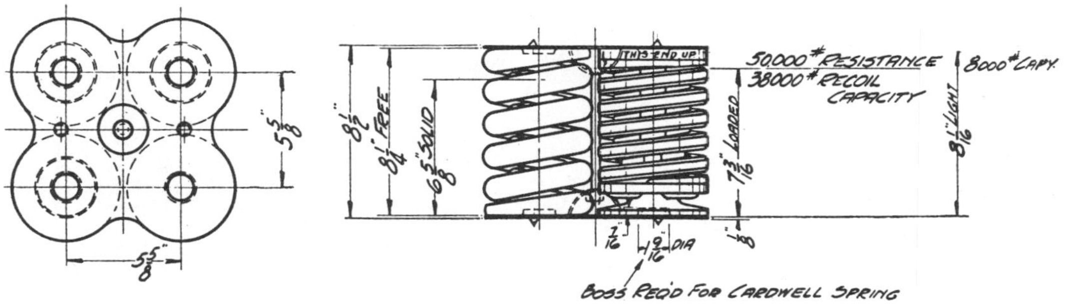

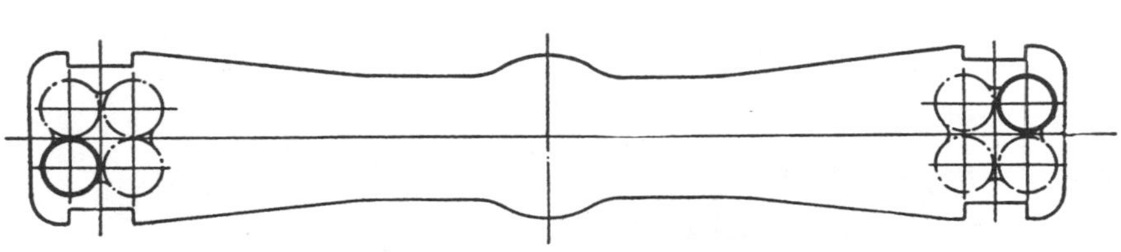

Fig. 14.163—(Left) Recommended location of Cardwell Friction Springs under the bolster ends.

Cardwell Westinghouse Company

(*See also Page* 1153)

Cardwell Friction Bolster Spring—Type A

Smoother riding for light as well as loaded cars

Eliminates derailments caused by excessive recoil on rough track

THE A. A. R. Bolster Springs recoil with approximately the same force as the force of the shock causing them to be compressed, resulting in a periodical vibration of compression and recoil for as many as thirty times before the car body is returned to its natural riding line.

To prevent this excessive compression and recoil action, with the resulting bouncing and rocking of car bodies, the use of a frictional snubbing element in bolster spring groups is now considered advisable for all cars in high speed service.

Cardwell Friction Bolster Spring Type A

The Cardwell Friction Bolster Spring provides for the higher operating speeds of freight cars, with safety, and smoother riding qualities with loaded as well as light cars.

In addition to reducing the recoil force of the spring nest, and increasing the spring capacity, the Cardwell Friction Bolster Spring prevents unnecessary spring movements and protects the A. A. R. coils from being driven solid, thereby greatly increasing the life of the A. A. R. Springs.

The truck column wear resulting from bolster movements is also greatly reduced and the car parts in general will show increased life with a reduced maintenance.

The application of the Cardwell Friction Bolster Spring is made by removing one of the A. A. R. Bolster Springs per nest and replacing same with Cardwell units. These units fit into the same space occupied by the A. A. R. coils removed.

Installation in 50-Ton Double Truss Truck
A.A.R. 1936 Springs

All wearing parts of the Cardwell Friction Bolster Spring are made of hardened steel under careful supervision, which insures long wearing life.

If further information is desired, this manufacturer will be pleased to have one of his representatives call and discuss individual requirements and furnish drawings showing recommended applications.

CARDWELL WESTINGHOUSE CO., CHICAGO

CANADIAN CARDWELL CO., LTD., MONTREAL

Other Products and Branch Offices Are Listed in the Classified Indexes

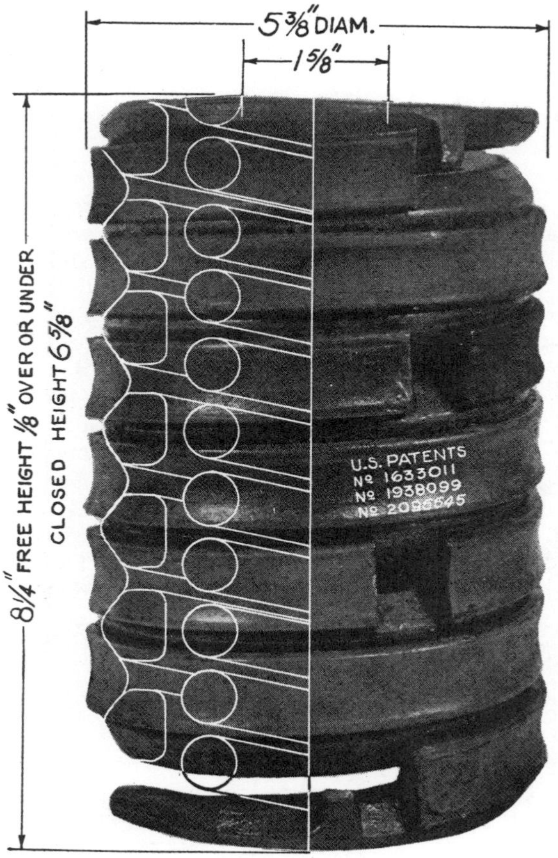

Fig. 14.164—Frost friction truck spring No. 360.

Fig. 14.165—Frost friction truck spring applied in standard A.A.R. four spring nest.

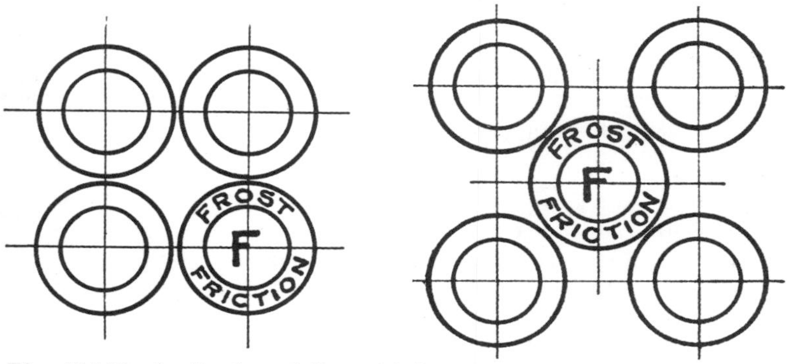

Fig. 14.166—Application of Frost friction truck spring to four-spring nest and five-spring nest.

The Frost Railway Supply Company

(*See also Page* 1155)

Fig. 14.167—Passenger car truck on C. M. St. P. & P. with coil-spring suspension, Monroe hydraulic shock absorbers and levelizing bars.

Description: *Railway Mechanical Engineer,* March 1939

Fig. 14.168—Miner Class C-2-XB Spring Snubber assembled in a four-wheel spring cluster.

W. H. Miner, Inc.

(*See also Page* 1162)

Frost Friction Truck Spring No. 360

Prevents damage to track, car, and lading due to harmonic vibratory action
Designed to replace one helical spring in standard spring groups

THE destructive effects of vertical harmonic action generated in freight cars moving at critical speeds can be avoided by installing a Frost Friction Snubbing Spring No. 360 in place of one helical spring in truck spring nests. In both laboratory and service tests this spring has demonstrated its capacity to eliminate the build-up of excessive and violent oscillation or bounce caused by the harmonic vibratory action characteristic of uncontrolled helical truck springs when subjected to impacts resulting from irregularities of track. Such oscillations are seriously injurious to car, lading, and roadbed. Therefore, this Frost Snubber provides a ready effective way to make freight cars give longer service and at the same time promote safety of operation while reducing expenditures for both car and track maintenance. Neither does the general design of the truck require modification in any way in order to reap these benefits.

Age in Months	Approximate Miles in Service	Kind of Service	Absorption in Foot Pounds under Movement of ¾ in.	1 in.	Absorption in per cent—New Spring taken as 100% ¾ in.	1 in.
I	II	III	IV	V	VI	VII
1	530	Coal Car.......	425	485	100%	100%
15	50,000	Refrigerator....	542	612	128%	126%
28	250,000	Loco. Tender...	248	308	57.0%	63.5%

Fig. A—Energy absorption of 40-ton spring groups containing one Frost Friction Truck Spring No. 360 in each group

The Frost Friction Truck Spring is designed to replace one coil spring in each standard helical truck spring group. It is applied either end up because it is symmetrical. It is a replacement, not an addition—the total weight of spring steel per car is not increased.

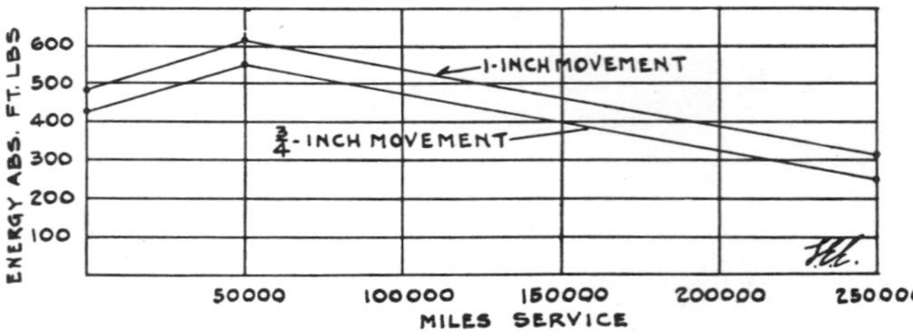

Fig. B—Energy absorption plotted against length of service

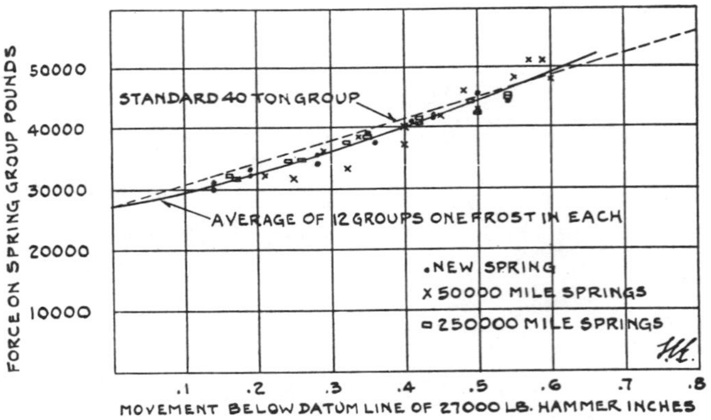

Fig. C—Impact pressures—40-ton spring groups containing one Frost No. 360 spring each, compared to standard 40-ton group

Spring Energy Absorbed by Friction

This Frost Friction Truck Spring, a snubbing spring, accomplishes its purpose by giving the groups of helical springs the necessary friction or energy absorption capacity to reduce materially, excessive and violent oscillations, thus reducing the number and intensity of overload blows. Tests have shown that with a Frost Snubber Spring in the group the nest of springs never goes solid — it absorbs energy in proportion to the load upon the spring.

This Snubber Spring is made of heat - treated spring steel. The total area of all friction surfaces is 140 sq. in. This large friction surface, being hardened spring steel, actually polishes in service and give a long service life.

Results of Tests

An accompanying Table, Fig. A, gives the results of test made under the 27,000-pound hammer at Purdue University, the figures showing the absorption in foot-pounds, each line being the average of four groups of springs in 40-ton group of 1915 standard springs with one Frost Friction Truck Spring No. 360 in each group.

The chart in Fig. B shows the average absorption of 40-ton group with different length of service Frost Springs in each, as compared to the absorption of 50 foot-pounds for ¾-in. movement, and 75 foot pounds for 1-in. movement for the 40-ton A. A. R. coil spring group.

The average pressure produced by the 40-ton group, one Frost Spring in each is shown in Fig. C. This chart gives the force of blows on the spring group as the hammer is raised above the datum line and then allowed to drop. These results indicate that the Frost No. 360 Friction Spring does its work of snubbing vibrations with no extra pressure in excess of that of an all coil spring group.

THE FROST RAILWAY SUPPLY CO., DETROIT, MICH.

Branch Offices Are Listed in the Classified Indexes

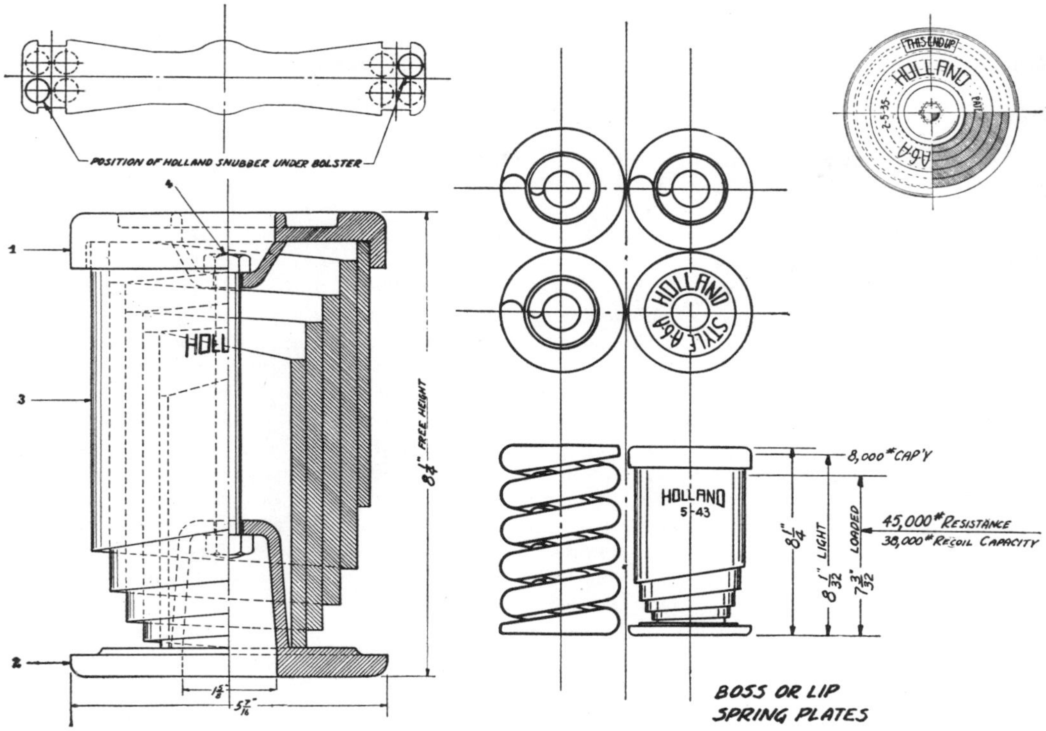

Fig. 14.170—Application of Holland Friction Volute Snubber Spring, Style A-6-A.

Fig. 14.171 — Volute Friction Spring.

Fig. 14.173—Upper Spring Seat.

Fig. 14.172—Bolt and Nut.

Fig. 14.174—Lower Spring Seat.

Holland Company

(*See also Page* 1157)

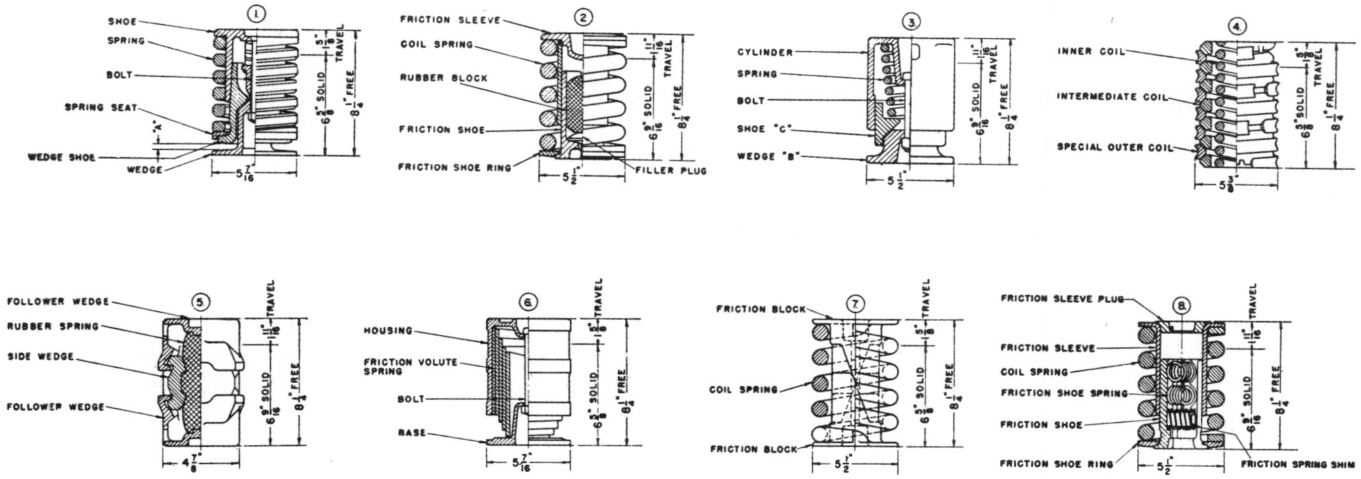

1—Cardwell Type A
2—Symington-Gould SBR
3—Miner C-2-XB
4—Frost 360
5—American Steel Foundries Simplex
6—Holland Style A-6-A
7—Gustin-Bacon A-1
8—Symington-Gould SBS

Fig. 14.174A—Proper method of applying various makes of Spring Snubbers.

(*A. A. R. Circular, May* 15, 1943)

Section 16

Industrial and Mine Cars

INDUSTRIAL cars are used in almost every branch of industrial activity; in manufacturing plants, mines, lumber operations, sugar cane plantation, and in construction work. They are also used in quarries, sand and gravel plants, brick and clay works, steel mills, smelters, foundries, machine shops and in many other places for handling all classes of loose or package materials. The classes of cars referred to as industrial include many types which often are also used in general railway service, and other smaller cars which are purely industrial.

These cars may be classified according to the industrial field in which they are used and according to their general form; dump cars, platform cars, hopper-bottom cars and various forms of rack cars.

Dump Cars

Cars of the dump type are adapted for handling a wide range of materials, and, because of the steep angle of discharge generally obtained, wet as well as dry materials may be dumped successfully. Cars of this type are used in mining and quarrying works; at sand and gravel plants; stone crushers; cement mills; fertilizer, phosphate and nitrate plants; powder works; smelters; clay, pottery, brick and glass works; by contractors for road paving, excavating and concrete work; in foundries for sand, slag and refuse and in chemical and various other industrial plants, particularly those using loose bulk materials.

Dump cars are made in various sizes and shapes to meet many requirements but differ considerably in details of design. The two general types are the box-body type and the V-body type; there are many modifications of both types. Generally, these cars are designed to dump toward either side, but they are also built so that the load may be discharged from the end.

Cars of the box-body type are used largely in construction work. They are generally referred to simply as "dump cars" although some of them are sometimes called "two-way" dump cars to distinguish them from "one-way" dump cars. In addition to the construction field, where excavating, filling and ballasting work is done, they are also employed around ore reduction plants, stone quarries and rock crushers, gravel and sand pits. They are useful for moving practically all classes of loose bulk materials and also such materials as pig iron, castings, small forgings, steel billets, scrap materials, etc.

The V-body dump car is generally built entirely of metal and to dump toward either side. It is particularly useful in industrial plants and for construction operations, because of its adaptability for handling any class of loose materials such as coal, ashes, sand, gravel, shale, rock, ore, earth, cinders, cement, concrete, clay, and other similar loose bulk material.

Another type of dump car is the rotary type which has either a square box or a scoop type of body which can be rotated and dumped in any direction.

Hopper Bottom Cars

Hopper bottom cars are extensively employed for handling coal, ore and loose bulk materials such as ashes, coke, sand, gravel, crushed stone, earth, clay, fertilizer, etc. They have also been adapted for use at manufacturing plants, power houses, fertilizer works, coal and ore mines, and in refining operations.

Gable Bottom Cars

Gable bottom cars are particularly adapted for use on inclined and elevated tracks or trestles. The facility with which the load may be dumped to either or both sides makes them an efficient and economical type of car for delivering many kinds of loose bulk material to storage bins. They are used at coal yards and power houses; at mines for handling iron and copper ore; at brick yards for handling clay; at sand and gravel pits, and for many other similar purposes. With slight modifications they are used with automatic and cable railways employed on the water front for handling coal from vessels and barges to storage bins and pockets.

Inclined Bottom Cars

Cars of the hopper-bottom type which discharge in the center, or of the gable-bottom type which discharge on the sides, meet many requirements but there are some places where inclined bottom cars which discharge on one side only, or at one end only, give more satisfactory service. Such cars serve to handle practically any class of loose material and are built in many different designs.

Platform Cars

The platform car, or "flat" car as it is generally termed, consists of a frame or flat body mounted on wheels and is used for general utility purposes at industrial plants or elsewhere. Service conditions vary widely and many variations in dimensions, capacity and details of designs are available; these cars are built in large eight-wheel units and also in many designs of small four-wheel cars.

Logging Equipment

Cars used for logging work usually consist of either a flat car or of two four-wheel trucks fitted with special types of log bunks. When trucks are used, the logs rest on the special bolster which forms the log bunk. The two trucks are sometimes connected together by a coupling beam, but frequently the logs form the only connection between the trucks. On logging cars designed for hauling long logs, no platform is placed on the frame and the center sills are sometimes fitted with an adjustable section so that the distance between the truck centers can be adjusted to suit the length of the logs.

Other Types

In addition to the types described there are many other cars designed for a wide variety of uses. These include motor operated cars of many kinds; charging cars for foundry work; dryer cars of the "rack" design for use in brick manufacture, core oven work, and special cars for creosoting operations.

Many dump cars also used in industrial operations are illustrated in Section 4.

Cane and other types of industrial cars used in other countries will be found in Section 18, Foreign Cars.

Fig. 16.01—Coal Mine Car.

Fig. 16.02—End Dumping, Cable-Hauled Metal Mine Car.

Fig. 16.03—Side-Dumping Metal Mine Car.

Fig. 16.04—Side-Dump Metal Mine Car for Inland Steel Company.

Fig. 16.05—Welded Steel "Gunboat" Type Coal Mine Car for Philadelphia & Reading Coal & Iron Works. Capacity 340 cu. ft.

Fig. 16.06—End-gate Coal Mine Car for Lehigh Navigation Coal Co. Capacity 114.8 cu. ft.

Fig. 16.07—Eight-wheel Large-capacity Coal Mine Car for Rochester and Pittsburgh Coal Company.

Fig. 16.08—Coal Mine Car for Inland Steel Company.

Fig. 16.09—Coal Mine Car for Amherst Coal Company.

American Car and Foundry Company

(*See also Page* 1189)

a.c.f. Industrial and Mine Cars and Parts

All types of equipment designed and built to a.c.f. or customer's specifications
Complete engineering and research supplement strategically located plants

a.c.f. has complete facilities for designing and building all types of industrial and mine cars. a.c.f. is also especially equipped to furnish repair parts, wheels and coupler rigging.

a.c.f. Mine and Industrial Cars

American Car and Foundry Company builds mine and industrial cars of all descriptions, to meet all practical requirements—of all steel, of composite steel and wood, or of all wood adequately ironed off with car forgings; automatic drop bottom cars as well as rotary dump cars or cars with end gates; high as well as low cars; small and large capacity.

Automatic Drop-Bottom Mine Car for different capacities and gages

Large capacity 8-wheel a.c.f. Mine Car for Rotary Dump to be used inside and outside the coal mines

Special attention is directed to the 8-wheel mine car illustrated on this page. This car was manufactured for a large company in the United States and is fitted with through axles having a.c.f. chilled tread wheels, equipped with modern anti-friction bearings in the wheels. As each wheel is free to rotate on its axle independently, there is no slipping and sliding of wheels, even on the shortest curves.

End Dump Mine Car built for Ford Motor Company Capacity, 122 cu. ft.

This mine car was designed for end dump unloading, but models for rotary dump and various track gages, are also available. Electrically welded end-sill construction and spring bumpers when desired.

The highly successful and favorably known a.c.f. electrically welded end-sill construction with built-in-spring bumper assures maximum strength and endurance. The drop-bottom doors are equipped with the a.c.f. simplified operating and locking mechanism. All moving parts, including wheels, bearings, and door control mechanism, are fully accessible.

a.c.f. Parts for Industrial and Mine Cars

a.c.f. Supply Service is ready at all times to furnish Chilled Tread Car Wheels, Chilled Tread Motor Wheels, Axles, Boxes, Car Forgings, Links and Pins, Couplings of all descriptions, Drawbars, Special Iron for Couplings and Drawbars, Standard Bar Iron, Special Bolts and Nuts, Cast Iron Flanged Pipe and Fittings.

Side-dump industrial car built for Pleasantville Constructors, Inc.

a.c.f. Research

a.c.f. engineers are backed by a research organization devoted to the problems of producing the best mine cars and wheels that modern engineering science can devise.

a.c.f. Plants

a.c.f. plants are within easy reach of the great industrial centers. These plants not only make cars, car parts and wheels, but they render a mine car engineering service unsurpassed in thoroughness.

The nearness of these big plants to coal properties and the field force representing each plant, make possible the close, personal contact which distinguishes a.c.f. service.

AMERICAN CAR AND FOUNDRY COMPANY, NEW YORK, N. Y.

Other Products and Branch Offices Are Listed in the Classified Indexes

Fig. 16.10—Eight-ton Double-truck Coal Mine Car.

Bethlehem Steel Company

Fig. 16.11—End-Dump Coal Mine Car for Leckie Collieries Co. Capacity 106 cu. ft.

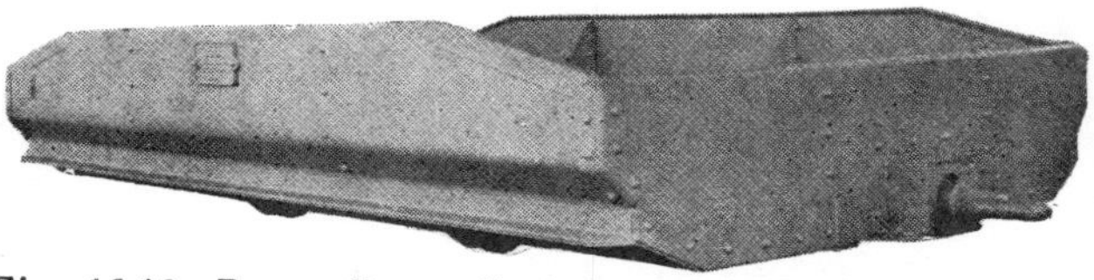

Fig. 16.12—Rotary-Dump Coal Car for Fifth Vein Coal Co. Capacity 171.5 cu. ft.

Figs. 16.13, 16.14—Side and bottom views of a large Four-Wheel Bottom Dump Mine Car for Snow Hill Coal Corp. Capacity 260 cu. ft. level.

Fig. 16.15—Alloy-Steel Coal Mine Car. Capacity 128.24 cu. ft. level.

Fig. 16.16—Two-Way Dump Car for Pleasantville Constructors, Inc. Capacity 5 cu. yds.

(*See also Page* 1189)

Fig. 16.17—Two-Way Side-Dump Car for Aluminum Company of America. Capacity 5 cu. yds. Bauxite.

Fig. 16.18—40-cu. ft. "V" Body Dump Car for Handling Coal.

American Car and Foundry Company

(*See also Page* 1189)

Fig. 16.19—Gun car—United States Army, Capacity 250 tons.

Fig. 16.20—All-steel, three-compartment, one-side-discharge Cane Car for United States Sugar Co. Capacity 60,000 lb., length over end sills 39 ft. 3⅞ in., track gage 4 ft. 8½ in.

Fig. 16.21—(Left) Triple V-body dump car, capacity 15 cu yds.

Fig. 16.22—(Below) All-steel logging car—Capacity 30 tons. Built for J. Neils Lumber Co.

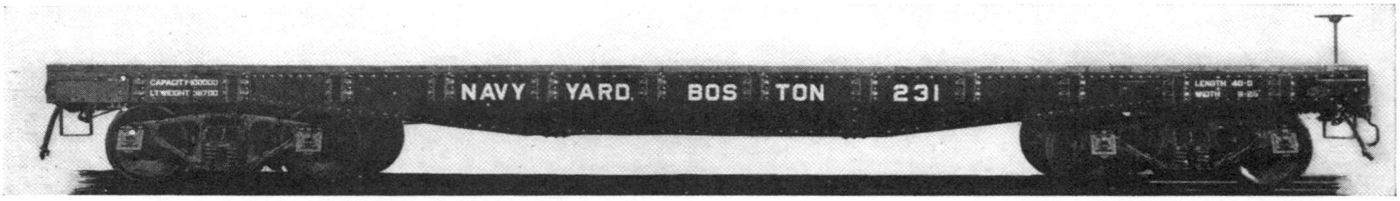

Fig. 16.23—Fifty-ton Flat Car, 40 ft. 0 in long, for U. S. Navy.
Magor Car Corporation

Fig. 16.25—50-ton all-steel automatic air dump car—Weight 77,300 lb.; capacity 810 cu. ft. or 30 cu. yds.

Fig. 16.26—70-ton all-steel automatic air dump car with down-folding doors; Capacity 810 cu. ft. or 30 cu. yds.

Fig. 16.27—Sand and gravel hopper car. Capacity 15 cu. yds.

Fig. 16.28—Ingoldsby type ore car. Capacity, 160,000 lb.—1,334 cu. ft.; light weight, 49,000 lb.

Magor Car Corporation

(*See also Page* 1193)

Air Dump and Industrial Cars by Magor and National

MAGOR Car Corporation and National Steel Car Corporation, Ltd., have engineering and plant facilities for the design and construction of all types of air dump and industrial cars.

The ore car built for The International Nickel Company, Inc., is a good example of the rugged, sturdy type of car required in mining operations.

40-Cubic Yard Twin Bodies Air Dump Car

30-Cubic Yard Air Dump Car

5-Cubic Yard Air Dump Car

Ore car built for The International Nickel Company, Inc.
Inside length, 22 ft. 11½ in.; cubic capacity 1086 cu. ft.

Air dump cars built by these Companies are in service all over the world and are noted everywhere for their economical handling of bulk materials. Illustrations on this page show typical examples of these cars. The complete range of capacities includes equipment from 5 cubic yards to 45 cubic yards.

Both Magor and National can furnish cars from standard designs or will prepare designs and estimates for particular projects. By consulting Magor and National it is possible to save on time, initial expenditures, and operating expense.

MAGOR CAR CORPORATION, NEW YORK, N. Y.
NATIONAL STEEL CAR CORPORATION, LTD., HAMILTON, ONTARIO, CANADA

Other Products and Branch Offices Are Listed in the Classified Indexes

Fig. 16.29—5 cu. yd. Heavy duty Granby type Ore Dump Car. 36-in. track gage.

Fig. 16.30—Granby type Ore Dump Car. Same as Fig. 16.29 in dumping position.

Built for Republic Iron & Steel Co.

Fig. 16.31—One-Side Dump, Mechanically Tilted Ore Car. Capacity 83 cu. ft., track gage 24 in.

Fig. 16.32—One-Side Dump Ore Car. Same as Fig. 16.31 in dumping position.

Built for Oliver Iron Ore Co.

Fig. 16.33—Rocker Dump Car for heavy steel mill service. Equipped with lever dumping device. Capacity 4 cu. yd., track gage 36 in.

Fig. 16.34—Rocker Dump Car. Same as Fig. 16.33 in dumping position.

Built for Republic Iron & Steel Co.

Pressed Steel Car Company, Inc. (Industrial Division)

Industrial and Mine Equipment

All steel and composite cars designed and built for mining, manufacturing plants, quarrying, and contracting

PRESSED Steel Car Company, Inc., has a complete engineering staff and modern plant facilities for designing and building all types of industrial, mining, and quarry cars, also individual cars for special service such as tunneling operations and the handling of special types of lading.

Cars for Mine Work

The coal car illustrated at the right is an example of cars built for coal mine operation. This car is of all-steel construction, ruggedly built to stand up under the most severe conditions with minimum attention for maintenance and repairs. Pressed Steel is always ready to submit estimates and designs for any type of mining operation.

Cars for Quarrying

Illustrated in the center is a modern type of car designed for use in quarrying. Pressed Steel has built large numbers of cars for this industry. Included are combination steel and wood as well as all-steel cars, hand, power, or air dumped, either standard or special designs. Capacities range from ⅔ cu. yd. to 30 cu. yd.

B—All-steel Tray type Quarry Car. Capacity 4½ cu. yd. Track gauge, 36 in. with hand brake. Car body is mechanically tilted to dump. Blair Limestone Co.

Special Cars for Handling Cable Reels

The car illustrated below is an excellent example of an industrial car built for a highly specialized pur-

A—End-dump Mine Car with link and pin type couplers and roller bearings. Vesta Coal Co.

pose. Pressed Steel Car also builds special types of cars for increasing operating efficiency of steel mills, manufacturing plants, and lumbering.

Other Types of Cars

Illustrated on adjoining pages will be found other examples of industrial and mine cars designed and built by this Company.

With its large plants, located in the Pittsburgh and Chicago Districts, Pressed Steel Car Company, Inc., is in a position to build all types and sizes of cars, both riveted and welded construction, including hopper cars, flat bottom gondolas and general service type; steel and composite box and automobile cars; ore cars; automatic air dump cars; caboose cars; and sugar cane cars.

C—Flat car with "fish belly" sills, 100 tons capacity, for handling large cable reels.

Fig. 16.35—All-steel heavy-duty quarry car.

Fig. 16.36—90 cu. ft. capacity Hopper Bottom Car for handling pulverized material. Aluminum Company of America.

Fig. 16.37—One-side-discharge car for handling petroleum coke.

Fig. 16.38—Gable bottom ore car for use in copper mining operations.

Fig. 16.39—All-steel, welded, Side Discharging Car for handling wet powder. Gate in open position.

Fig. 16.40—Standard all-around dump car. Shown in dumping position.

Pressed Steel Car Company, Inc. (Industrial Division)

(*See also Page* 1195)

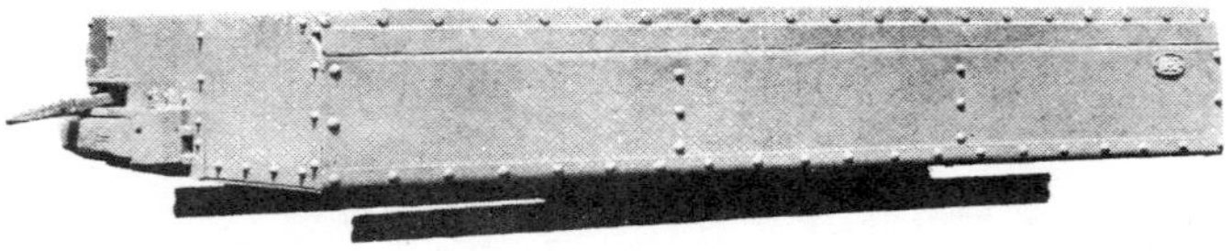

Fig. 16.41—Low-height, rotary-dump Mine Car with spring bumpers and roller bearings.

Fig. 16.42—Koppel 5-ton capacity Platform Car with roller bearings and wood or steel platform.

Fig. 16.43—Heavy-duty one-side-dump Quarry Car. Capacity 4 cu. yds.; 36 in. track gage. Universal hand brake.

Fig. 16.44—Large-capacity rotary-dump type Mine Car with tight-lock swivel type couplers. Pittsburgh Coal Co.

Fig. 16.45—135-ton Transformer Transfer Car.

Fig. 16.46—One-side-dump Quarry Car. Same as Fig. 16.43 in dumping position. Body tilted by overhead dumping arrangement or device parallelling the track.

Fig. 16.47—Motor-driven Rocker Car with body in dumped position. Capacity 2 cu. yds., track gage 36 in. Solenoid brake and cable reel mounting. U. S. Gypsum Co.

Fig. 16.48—Standard gage Triple Cinder Car of 100-tons capacity, Wheeling Steel Corp.

Also designed to accommodate one or two ladles.

Pressed Steel Car Company, Inc. (Industrial Division)

(*See also Page* 1195)

Fig. 16.49—Two-way Dump-Car equipped with scale. U. S. Engineers.

Fig. 16.50—Cradle Dump Car with adjustable dumping arrangement to hold body in semi-dumped position.

Fig. 16.51—One-Side-Discharge Eight-Wheel Car. Mechanically tilted to dump. Traux-Traer Coal Co.

Fig. 16.52—Rigid-Body End-Dump Car with swivelling automatic couplers.

Fig. 16.53—One-Side-Discharge Dump Car — Manually operated—Automatic couplers.

Fig. 16.54—Low-Height, Two-Way Dump Car which discharges to either side.

Fig. 16.55—Automatic 50-ton standard gage Drop Bottom Car. Doors released and closed when car is in motion by a tripping device parallel to the track.

Pressed Steel Car Company, Inc. (Industrial Division)

(*See also Pages* 1195, 1199)

Air Dump Equipment Built by Pressed Steel Car

Cubic capacities, 12 to 50 cu. yd.—Load capacities to 100 tons
Quick dumping—Ruggedly constructed—Highly dependable

PRESSED Steel Car Company, Inc., Industrial Division, specializes in the design and manufacture of cars for the mining industry and the industrial field generally. Many years of development work in these fields has resulted in a wide range of standard and special designs to meet almost every possible requirement of material handling.

This page, dealing only with cars of the air dump class, shows views of cars representing two distinct methods of operation. In the one case the cylinders are attached directly to the car, permitting the car to be dumped wherever desired, either individually or two or more cars simultaneously. In the other, the cars are dumped individually at a fixed point, such as at the crusher house, and in an operation of this kind, the cars are not equipped with air dump cylinders, but suitable cylinders are installed at the point where the cars are dumped. This arrangement has the advantages of lower first cost of cars and lower maintenance cost, as compared with cars equipped with dumping cylinders, and has worked out very satisfactorily in a number of mining operations over a period of some years.

For other types and sizes of cars, see pages 1194 to 1198 inclusive.

The above views show cars equipped with air dump cylinders of the heavy duty type, 30 cubic yards level, 75 tons load carrying capacity. Cars of this type are furnished in a range of capacities from 12 to 50 cubic yards level.

These three views below show cars of the air dump type, 30 cubic yards level capacity, but not equipped with air dump cylinders. Two cylinders are installed in the crusher house. The cars are dumped by means of these cylinders individually but quickly into the hopper leading to the crusher. When cars of smaller capacity are used, one dumping cylinder is ample. This method of operation is very simple and has proved exceedingly economical for cars ranging in size from 12 to 30 cubic yards level capacity.

Left:
Full-dump position, bottom view. Note absence of dump cylinders

Right:
Full-dump position, top view. Note free clearance for discharge of load

PRESSED STEEL CAR COMPANY, INC., Industrial Division, PITTSBURGH, PA.

Other Products and Branch Offices Are Listed in the Classified Indexes

Fig. 16.56—50-ton bloom car. Built for Monessen Southwestern Railway by Pressed Steel Car Company, Inc. (Industrial Division).

Fig. 16.57—Special Stake Car for steel mill service. Capacity 90 tons. Built by General American Transportation Corporation.

Fig. 16.58—Side dump slag car for Noranda Mines, Ltd. Builder, Canadian Car & Foundry Company, Ltd.

Fig. 16.59—End dump slag car for Noranda Mines, Ltd. Builder, Canadian Car & Foundry Company, Ltd.

Fig. 16.60—Hot metal car for transporting molten iron a distance of 11 miles from blast furnace to open hearth furnace. The American Rolling Mill Company, Middletown, Ohio. Length coupled 56 ft. 6 in.; truck centers 30 ft. 3 in.; total wheel base 45 ft. 6 in. Loaded Weight 685,000 lb. Builder, M. H. Treadwell Company.

Section 17

Export Cars

Freight and Passenger

THE USE OF AMERICAN BUILT RAILWAY EQUIPMENT has become world-wide on foreign railroads and in industrial operations. Therefore, this class of equipment is included in this book so that information on the entire scope of American carbuilding may be available. The widely varying conditions encountered in freight and passenger service in foreign lands has necessitated the construction of many different types of cars; in some countries—particularly for European freight traffic—small cars of the four-wheel type predominate while in other countries—Cuba, Mexico, South America, South Africa, Australia, China and other parts of the Orient—eight-wheel or double-truck cars, more like those on American railways, are extensively used in main line service.

CONTENTS OF SECTION 17

Passenger Cars 1202

Freight Cars 1204

For Additional Export Industrial and Mining Cars see Section 16

Foreign Passenger Cars

Passenger cars used on the railways throughout the world range from the very simply constructed wooden third-class car to large, finely appointed steel equipment which compares favorably with the best types of passenger cars in use on American railways.

In European countries and many parts of the British Empire passenger cars of the side-door compartment types are in general use, but in other parts of the world, particularly in Latin America, the end-door type, similar to American railway equipment, is extensively used. In many countries the railroads operate first class, second class, and third class passenger coaches; many of the progressive railways have dining and sleeping cars, and also baggage, express, and other types of passenger train equipment similar to those used on American railways.

Generally these cars are equipped with American types of couplings and brakes but European and special types of appliances are also provided when necessary to conform to the practices already established on the railway for which the cars are designed.

Foreign Freight Cars

The freight cars are built in both the four-wheel and eight wheel types and include box cars, gondolas, flat, hopper, tank, and other types. Many of the lighter cars are provided only with hand brakes but the use of power brakes—vacuum type or some other type of air brake—is quite extensive on the heavier equipment.

The screw type of coupler, supplemented by side buffers, is in extensive use in European countries; in many other countries the American type of automatic coupler or a modification of it is used; the link-and-pin coupler, and various designs of hooks, springs, etc., are also used on much of the light equipment.

The four-wheel cars have the axles fixed to the car frame and are provided with springs—usually semi-elliptical, sometimes coil—over the axles. Some of these cars are of the closed-top type but in many cases they are simply small gondolas or open-top cars, provision being made to cover, with a tarpaulin, goods which are liable to damage from exposure to the weather.

The heavier eight-wheel cars are constructed along lines more closely following American practice and are equipped with many of the same appliances—draft gear, automatic couplers, air brakes, etc.

Cane Cars

Sugar cane plantations are operated in the United States, in Cuba, Porto Rico, Mexico, Central and South America, Hawàii, and other tropical countries. This has resulted in the development of many different types of cane cars to meet special requirements.

These cars are of two general groups; large double-truck or eight-wheel cars; and small four-wheel cars. They may be of the side discharge type, end discharge—not extensively used; or designed for hoist discharge or rake discharge. In the largest cane producing sections, especially in Cuba, car dumpers of the side tipping platform type are extensively used with side discharging type of car.

Four-wheel cane cars differ widely in design and are used extensively. With little change in details they are used on other than cane plantations for handling such material as sisal grass, henequen, manila, bananas, etc.

Fig. 17.01—Mixed-Class Coach for Mozambique Railway, Portuguese East Africa.
Description: *Railway Age,* July 11, 1942

Fig. 17.02—Second-Class Compartment in Coach.

Fig. 17.03—Third-Class Compartment.

Fig. 17.04—Brake End of Mixed Coach.

Fig. 17.05—Brake End of Third-Class Coach.

American Car and Foundry Export Company

(*See also Page* 1209)

Fig. 17.06—Narrow-Gage Motor and Trailer Car for Colombian National Railways, South America.

Description: *Railway Age,* October 25, 1941

Fig. 17.07—(Right) Interior of DeLuxe Motor Car.

Service Weight Motor Car, 52,540 lb.
Service Weight, Trailer Car, 31,350 lb.

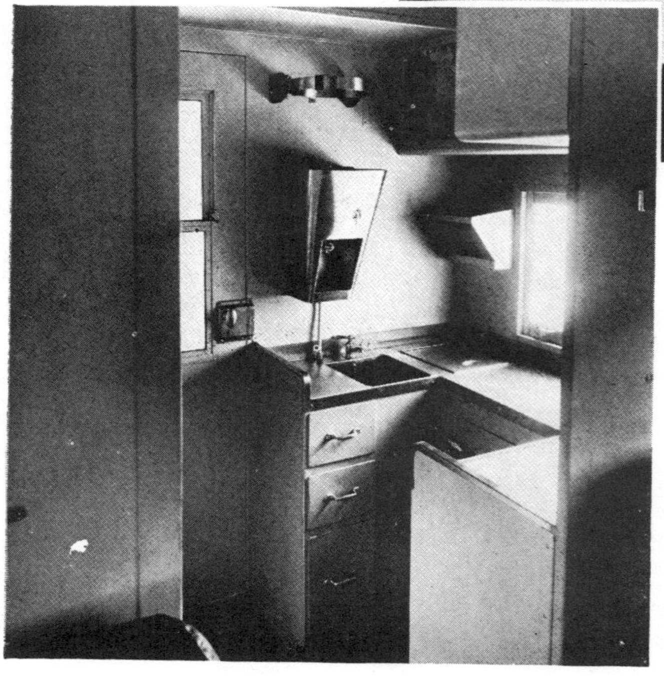

Fig. 17.08—(Left) Service Bar in motor for refreshments.

Fig. 17.09—(Below) 275-hp. Oil Engine.

American Car and Foundry Export Company

(*See also Page* 1209)

Fig. 17.10—Box car—Paulista Railways, Brazil. Capacity 42 metric tons, track gage 5 ft. 3 in.

Fig. 17.11 — Ore Car, Victoria and Minas Railway, Brazil. Capacity 50 metric tons, track gage one meter.

Fig. 17.12—Restaurant car—Rio Grande do Sul Railways, Brazil.

Fig. 17.13—Third-Class Coach, Portuguese Government Railways, Mozambique, Portuguese East Africa. Length over end sills 59 ft. 5¾ in., track gage 42 in.

Builder—**Pullman-Standard Car Manufacturing Company**

Exported by—**Pullman-Standard Car Export Corporation**

Cars Built by Pullman-Standard for Export

Passenger—Freight—Subway and Interurban

THE experience, equipment, and organization which have been developed by Pullman-Standard Car Manufacturing Company and Pullman-Standard Car Export Corporation over many years in this particular field enable these companies to meet the car requirements outside the United States more economically than can be accomplished with the facilities at the command of the average railroad.

The Companies mentioned above include among their assets engineering staffs, plants, and shipping facilities for promptly handling estimates and orders for all types of freight cars, also passenger, interurban and subway cars.

Illustrated on this page are examples of equipment built for foreign railroads. These cars have been selected from many similar designs and are representative of the capacity of these Companies, which completely covers the entire field of steel car construction.

All Steel Gondola Car. Capacity—60 metric tons. Track Gage—5 ft. 3 in. Built for Central Railways of Brazil.

Banana Car. Capacity, 30,000 lb.; Track Gage, 42 in. Built for Cia Bananera de Costa Rica

Multiple-Unit 3-Car Electric Train Built for Sorocabana Railway of Brazil

PULLMAN-STANDARD CAR EXPORT CORPORATION, NEW YORK, U.S.A.

EXPORTERS OF PRODUCTS OF

PULLMAN-STANDARD CAR MANUFACTURING COMPANY

Other Products and Branch Offices Are Listed in the Classified Indexes

Fig. 17.14—Box Car for South America, built by American Car and Foundry Co. and Pullman-Standard Car Manufacturing Co. Wolmanized lumber treated by the American Lumber & Treating Co.

Description: *Railway Age*, October 17, 1942
(*See also Page* 1238)

Fig. 17.15—Sleeping car—Rio Grande do Sul Railways, Brazil.
Builder—Pullman-Standard Car Manufacturing Company
Exported by—Pullman-Standard Car Export Corporation

Fig. 17.16—Stainless steel coach built for Companhia dos Caminhos de Ferro Portugueses, Portugal. Builder, Edward G. Budd Manufacturing Company.

Fig. 17.17—Flat Car for Central Railway of Brazil. Capacity 30,000 kilograms, lightweight 10,890 kilograms, track gage 5 ft. 3 in.

Fig. 17.18—Cane Car for Great Western Railway of Brazil. Three compartments, capacity 20,000 kilograms, lightweight 8,300 kilograms, track gage one meter.

Fig. 17.19—Box Car for Portuguese East African Railways. Two side doors. Track gage 42 in.

Fig. 17.20—Box Car for Great Western Railway of Brazil. Capacity 25,000 kilograms, track gage one meter.

American Car and Foundry Export Company

Fig. 17.21 — High-side gondola car for Central Railroad of Brazil equipped with hinged two-piece side doors for manual unloading.

Track gage, 1 meter
Light-weight, 13,970 kilos (30,780 lb.)
Capacity, 30,000 kilos (66,100 lb.)
Length inside, 34 ft. 5⅜ in.

Fig. 17.22—Side-unloading gondola car for Colombian National Railways, Colombia. One side door dropped down for unloading.

Fig. 17.23 — Small gondola car for Portuguese East Africa Railways.

Fig. 17.24—Steel box car for Nacionales de Mexico, Railways, Mexico.

Capacity, 50 tons
Light-weight, 44,500 lb.
Standard track gage

a.c.f Designs and Manufactures Cars for Export

Well known American builder is equipped to meet requirements of foreign countries

Product includes all types, capacities, and gages

THE experience of a.c.f in the design and construction of all types of railway cars for service outside the boundaries of the United States covers a period of almost a half century. This experience has brought a.c.f engineers into intimate contact with the practical railway operating problems as encountered in all parts of the world.

To solve these and kindred problems most efficiently and economically requires not only a knowledge of the problems themselves, but also of the various materials and manufacturing processes which are now available to the car builder's art. In this respect also, a.c.f excels, because, in the course of building thousands of cars of all types and capacities for American railroads, the a.c.f organization has obtained first-hand knowledge as a result of which the proper technique to fit the individual problem can be followed.

Examples of Recent Construction

A few examples of equipment built recently are shown on this page. These cars are of all-steel construction but if, in certain cases, the use of wood is desirable, a.c.f is ready to submit designs embodying wood construction in conjunction with steel.

The coaches for the Mozambique Railways of Portuguese East Africa are notable for the fact that low-alloy, high-tensile steels were used throughout the framing and strength members resulting in a substantial saving of weight on these cars.

Engineering Service

a.c.f is ready at all times to collaborate with the officials of prospective customer roads, in the preparation of designs or where the problem may not have reached the design stage, to render opinions based on the complete data of a.c.f engineering experience.

All-steel Box Car built for the Central Railway of Brazil

Mixed-Class Coach built for the Mozambique Railway, Portuguese East Africa. Weight, 55,900 lb.; passenger capacity, 12 first-class, 32 second-class, 80 third-class, total 124.

Side Dump Gondola Car built for the Colombian National Railways. Capacity, thirty-five metric tons. Equipped with automatic coupler; safety chains; and air brakes.

AMERICAN CAR AND FOUNDRY EXPORT COMPANY, NEW YORK, N. Y.

Other Products and Branch Offices Are Listed in the Classified Indexes

Fig. 17.25—10-yard level full, 40,000 lb. capacity, dual side pivot, drop door, automatic air dump car. 36-in. gage track. International Railways of Central America.

Fig. 17.26—16-yard level full, 60,000 lb. capacity, electrically propelled, side pivot, drop door, air operated dump car. Andalusian Mining Company.

Fig. 17.27—16-yard level full, 42 in. gage track, 60,000 lb. capacity, wood bed, steel underframe, lift door, two way, air operated side dump car. Portuguese East Africa.

Fig. 17.28—3-yard, 24 in. gage track, all-steel, one-way-dump tunnel car. Hawaiian Islands.

"Western" Cars for Export

THE "Western" line includes drop door and lift door air dump cars ranging in capacity from 1½ to 50 yards; also all types of special cars to meet customers' specifications.

"Western" cars have been manufactured for more than half a century, and there are large numbers in use today in all parts of the world. Their unexcelled reputation for efficient, low cost, reliable service adapts them particularly to work under the conditions prevailing in foreign countries.

The "Western" engineering department is available for consultation at all times, and will be glad to recommend the size and style of car best suited to any individual construction program.

30-Yard level full, 100,000-lb. capacity, heavy duty, one-way side pivot, drop door, air operated dump car with high clearance trucks for iron mining. Luossavaara Kirunavaara Co., Kiruna, Sweden.

22-Yard, 80,000-lb. capacity, all-steel, dual side pivot, drop door, automatic air dump car, 42-in. guage track. Union Miniere du Haut Katanga, Belgian Congo, Africa.

8-Yard, all-steel, two-way side dump, lift door dump car. 36-in. gauge track. Cuba & Central America.

30-Yard level full, 50-ton, dual side pivot, drop door, automatic air dump car with special sectional heat resisting floor for handling hot materials. 63-in. track gauge. National Steel Company of Brazil (Cia. Siderurgica., Nacional)

Fig. 17.29—All-steel heavy Quarry Car for a South African quarry. Capacity 8 cu. yds.

Fig. 17.30—Cradle Dump Car.

Fig. 17.31—All-steel Square-box Dump Car for steel plant, South Africa. Capacity 8 cu. yds.; roller bearings in wheel hubs.

Fig. 17.32—Small cane car with latticed endracks. Capacity 2 to 3 tons.

Fig. 17.33—Trunion type Air Dump Car with down-turning doors for Soviet Government. Capacity 100,000 lb., 30 cu. yds., track gage 5 ft. 0 in.

Pressed Steel Car Company, Inc. (Export Division)

Cars for Foreign Railroads and Industries

Designs cover wide range of capacities, types and gauges
Spare parts service also available

RAILROAD and industrial cars built and exported by Pressed Steel Car Company, Inc., are in use in many parts of the world. The experience of this Company in meeting the requirements of foreign service in different countries dates back to 1899. During that period railway car equipment in a wide variety of designs has been built for regular railroad operation, as well as for mining operations of all kinds, earth moving operations, manufacturing plants, sugar and sisal plantations and other industries.

Narrow gauge flat car of 15 tons capacity for the Belgian Congo, with Screw Couplings, Central Buffers and Screw Hand Brake

As a result of this wide experience, representatives and engineers of Pressed Steel Car Company, Inc. are in a position to offer constructive suggestions when the purchase of new equipment is contemplated.

Scope of Service Covered

The cars illustrated on this page are typical of the scope of the services where cars designed and built by Pressed Steel Car have been installed. On adjoining pages are other typical examples of equipment. If the particular type a customer may require is not shown, a request for further information will receive prompt and courteous attention and no obligation will be incurred.

Improved design balanced type all-around dump car for metal mining, with anti-friction roller bearings in wheel hubs and ball bearing turntable, 20 cubic feet capacity

Car Parts Can Be Supplied

Pressed Steel Car Company, Inc. not only designs and builds complete cars, but also is equipped to furnish spare parts of all kinds, including running gear with bearings and draft gear of various types; castings, forgings and pressed and welded members. Patented car parts are also available, as for example a safety link-and-pin coupler which avoids objectionable features ordinarily common to this type of coupling.

Heavy sugar cane car of 40 tons capacity for standard gauge main line haulage, with A.A.R. appliances, for Mexico. This is an unusually heavy car for this service

Plant Facilities

Pressed Steel Car Company, Inc. has large plants located in the Pittsburgh District and in the Chicago District of the United States and a large engineering staff is maintained, including specialists in particular industries. Plants are equipped with modern machinery for pressing, welding, and fabricating steel in all sizes and shapes.

PRESSED STEEL CAR COMPANY, INC., Export Division, PITTSBURGH, PA.

Other Products and Branch Offices Are Listed in the Classified Indexes

Fig. 17.35—Rocker dump ore car for mining operation in Bolivia. Capacity 18 cu. ft.; gage $19\frac{11}{16}$ in.

Fig. 17.36—Improved balanced type all-around-dump mine car with ball-bearing turntable and special features. Left—Car in dumped position. Right—End view showing control lever.

Fig. 17.37—Narrow width, low height, trunion type, two-way dump car. Equipped with lifting plates at each end for dumping body independently of underframe.

Fig. 17.38—Granby type Mine Car. Capacity three cu. yds., roller bearings in wheel hubs.

Fig. 17.39—All-steel Ventilated Box Car for Manila Railroad. 30 metric tons capacity.

Automatic couplers are provided with transition arrangement for service with non-automatic central hook couplings of European type.

Pressed Steel Car Company, Inc. (Export Division)

Special Car Parts for Export

Wheel and Bearing Assemblies—Link and Pin SAFETY Couplers for Industrial Cars

As ONE of the world's largest builders of Industrial Railway Cars, Pressed Steel Car has developed many special features in designing a wide variety of cars for every purpose and for greatest efficiency and safety in operation at the mine, mill, plantation or wherever work goes on.

Below are shown only a few typical running gear and draft gear assemblies. Many thousands of cars are in operation all over the world having these and many other special features. These are often adaptable to existing equipment. Special attention is given to repair and replacement parts service.

Typical Roller Bearing Assemblies for Inside and Outside Type Bearing Boxes

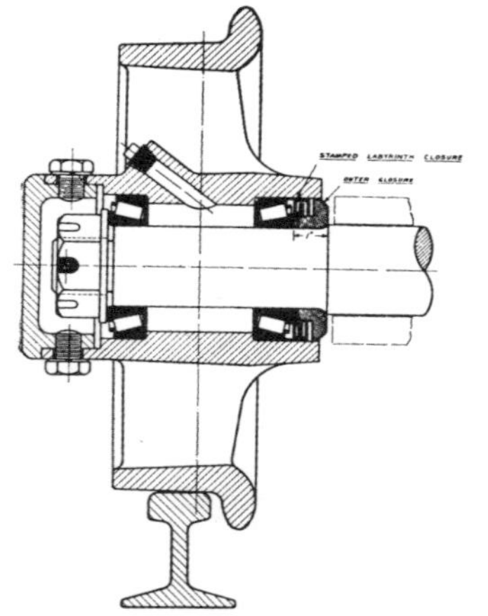

Wheel assembly with anti-friction Roller Bearings in hub, and improved type of labyrinth seal designed to protect bearing against dirt and water.

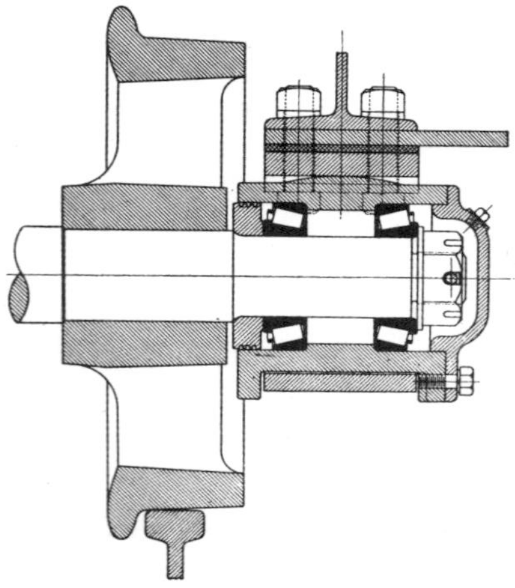

An outside type bearing box equipped with anti-friction bearings. These bearings are also suitable for inside type boxes.

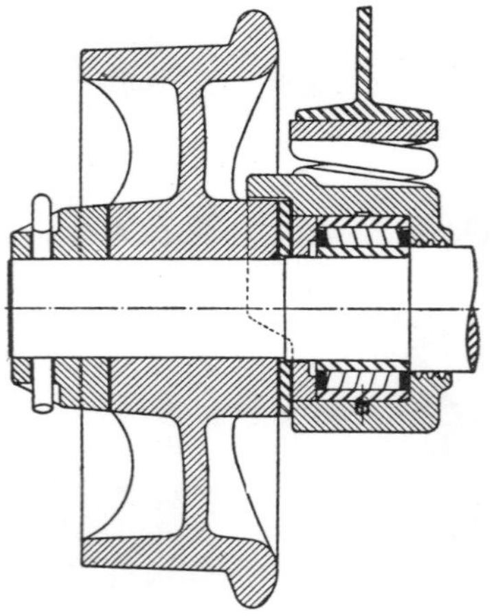

Inside type bearing box equipped with anti-friction roller bearings. These bearings are also suitable for outside type boxes.

Important Features of Link and Pin SAFETY Couplers with Retractable Link

Link in coupling position

Coupler with yoke

Link partly retracted

Link fully retracted

Coupler with end sill base

In operations where automatic couplers are not used and link and pin couplers are required, the above illustrations show a design which overcomes objectionable features inherent to the ordinary link and pin coupling. The link is fully retractable and non-removable, and is conveniently handled from underneath the coupler, permitting coupling with positive safety against injury to operator's hands. Cars may be coupled with coupler faces together on straight or curved track. Links lie in horizontal position when protruding from coupler, but are free to recede in coupler head without damage if contacted by the coupler of an approaching car. When coupler with end sill base is used, double coil springs are contained in the housing of the coupler to take care of buffing and pulling shocks. Link and pin and other parts are retained in coupler head and housing in such a manner as to prevent loss or convenient removal.

Wide and high buffer faces prevent interlocking or passing of coupler heads on curves or uneven track or when adjoining cars have somewhat different coupler heights.

PRESSED STEEL CAR COMPANY, INC., Export Division, PITTSBURGH, PA.

Other Products and Branch Offices Are Listed in the Classified Indexes

Fig. 17.41—Granby type mine car. 80 cu. ft. capacity.

Fig. 17.42—Ore car, 42 cu. ft. capacity, with swivel couplings and roller bearings. Dumped in rotary tipple.

Fig. 17.43—Heavy duty rocker dump car for copper ore in South Africa.

Fig. 17.44—Gable-bottom ore car of welded construction.

Fig. 17.45—Drop-Side Gondola Car for ballast and general service. Built for Venezuelan Government Railways.

Fig. 17.46—Logging car for Philippines. Capacity 40 tons.

Pressed Steel Car Company, Inc. (Export Division)

Fig. 17.48—Box Car for F. C. de Guantanimo and Occidente, Cuba. Capacity 40 tons; track gage 4 ft. 8½ in.

Fig. 17.49—Box Car sheathed with U.S.S. Cor-Ten steel for F. C. al Pacifico, Costa Rica. Length inside 35 ft. 0 in.; capacity 37 tons; track gage 42 in.

Fig. 17.50—Steel-frame, single-sheathed Box Car for Anglo Newfoundland Development Company, Newfoundland. Length inside 36 ft. 0¼ in.; capacity 30 tons; track gage 42 in.

Magor Car Corporation

Fig. 17.51—All-steel Air Dump Car for Bethlehem Chile Iron Mines Company, Chile. Capacity 30 cu. yds., track gage 4 ft. 8½ in.

Fig. 17.52—Steel Air Dump Car for Chile Exploration Company, Chile. Capacity 20 cu. yds., track gage 4 ft. 8½ in.

Fig. 17.53—Center-Hinge Type Automatic Air Dump Car with Apron, for Amtorg Trading Corp., Russia. Capacity 30 cu. yds., Weight 69,000 lb., track gage 5 ft. 0 in.

Fig. 17.54—Side-Hinge Type Automatic Air Dump Car for Amtorg Trading Corp., Russia. Capacity 20 cu. yds., Weight 44,000 lb., track gage one meter.

Magor Car Corporation

Freight, Work, Industrial, and Contractor Equipments by

Magor and National for Export

THE Magor Car Corporation and The National Steel Car Corporation, Ltd., have been actively engaged for many years in the export markets, and have supplied cars of all types to approximately 60 foreign countries scattered throughout the world. It is their vast experience in this field and their complete knowledge of requirements of foreign railroads that places them in a position to serve the export markets with efficient cars at economical prices. Into this ability enter careful methods of packing for ocean export, by which the largest possible savings in freight are coupled with the least possible costs in field erection at points of final destination. In order to give the best service to their numerous clientele, they have organized an Export Department and have appointed Agents in those countries where the demand for railroad equipment has warranted it.

Above:—20-ton, Steel-Frame Ballast Flat Car for Guatemala

Above: — 6 - Ton, All-Steel Cane Car for Venezuela

Above:—4-Wheel, "V"-type, Hand Dump Car for service in Jamaica—All-steel construction. Capacity, 13½ cu. ft.; gage 24 in.

Below:—4-Wheel, Equalizer Type Passenger Truck. Capacity, 40,000 lb.; gage, 42 in. For service in Costa Rica

Above:—Steel-Underframe Flat Car with End Racks. Capacity, 25,000 lb.; weight, 14,320 lb.; gage, 36 in. Used in Panama

MAGOR CAR CORPORATION, NEW YORK, N. Y.

NATIONAL STEEL CAR CORPORATION, LTD., HAMILTON, ONTARIO, CANADA

Other Products and Branch Offices Are Listed in the Classified Indexes

Freight, Work, Industrial, and Contractor Equipments by

Magor and National for Export

Above:—Steel Underframe Flat Car with Wood Floor. Capacity, 100,000 lb.; load limit, 128,000 lb.; weight, 41,100 lb. For service in Panama

Left:—Side-Hinge, Automatic Air Dump Car. Capacity, 20 cu. yd. or 30 tons; weight, 51,200 lb.; gage, 5 ft. 0 in. For service in Russia

Right:—16-cu. yd. U. S. S. Cor-Ten Steel-Lined Air Dump Car for Costa Rica

Below: — 10-ton, All-Steel Side Doors, Gable Bottom Gondola Car for Brazil

Above:—50-ton Ore Hopper Car for Chile

Composite, High-Side Gondola with Bottom Doors. Capacity, 100,000 lb. For service in Mexico

MAGOR CAR CORPORATION, NEW YORK, N. Y.

NATIONAL STEEL CAR CORPORATION, LTD., HAMILTON, ONTARIO, CANADA

Other Products and Branch Offices Are Listed in the Classified Indexes

Freight, Work, Industrial, and Contractor Equipments by

Magor and National for Export

Above—Right:—2-Compartment, one-side discharge, All-Steel Cane Car. Capacity, 10 metric tons; weight, 11,200 lb. For service in Brazil

Above:—20-ton Steel Frame Banana Car for Colombia

Below:—30-ton, Gable Bottom, Side Door Ore Gondola Car for Newfoundland

Above:—33-Ton, All-Steel, Side Dump Cane Car for Mexico

Below:—33-Ton, U. S. S. Cor-Ten Steel Banana Car for Nicaragua

MAGOR CAR CORPORATION, NEW YORK, N. Y.

NATIONAL STEEL CAR CORPORATION, LTD., HAMILTON, ONTARIO, CANADA

Other Products and Branch Offices Are Listed in the Classified Indexes

Freight, Work, Industrial, and Contractor Equipments by

Magor and National for Export

Left:—30-Ton, Steel-Frame, Steel-End Box Car for Mexico

Below:—All-Steel, 4-Wheel, High-Side Gondola. Capacity, 27,550 lb.; weight, 16,200 lb.; gage, 39⅜ in. For service in Thailand

Below:—All-Steel Tank Car. Capacity, 6,000 gal.; gage, 36 in. For service in Colombia

Below:—15-Ton, Steel-Frame Box Car for Venezuela

Above:—33-Ton, U. S. S. Cor-Ten Steel Box Car for Costa Rica

Right: — Steel Underframe, steel sheathed, wood-lined box car. Capacity, 100,000 lb. nominal; load limit, 123,500 cubic, 3712 cu. ft.; weight, 45,500 lb.

MAGOR CAR CORPORATION, NEW YORK, N. Y.

NATIONAL STEEL CAR CORPORATION, LTD., HAMILTON, ONTARIO, CANADA

Other Products and Branch Offices Are Listed in the Classified Indexes

Fig. 17.55—All-steel double deck Banana Car for Trujillo R. R., Honduras. Capacity 15 tons, track gage 36 in.

Fig. 17.56—Steel frame Cattle Car for F. C. Internacional de Central America, Guatemala. Capacity 20 tons, track gage 36 in.

Fig. 17.57—Steel frame, double-deck Banana Car for F. C. El Palito Palina Sala, Venezuela. Capacity 20 tons, track gage 42 in.

Magor Car Corporation

Fig. 17.58—Box Car for Sorocabana Railways, Brazil. Capacity 36 metric tons, track gage one meter. Exporter—Pullman-Standard Car Export Corporation

Fig. 17.59—Cane Car for Hacienda Laredo, Peru. Capacity 8 tons, track gage 42 in. Builder—Magor Car Corporation

Fig. 17.60—Steel frame, steel floor, Cane Car for Central Santa Marta, Cuba. Capacity 30 tons, track gage 4 ft. 8½ in. Builder—Magor Car Corporation

Fig. 17.61—Sand and Gravel Dump Car. Capacity 13½ cu. ft., track gage 24 in.

Fig. 17.62—All-steel Rock Car for Companhia Brasileira de Cimento Portland S. A., Brasil. Capacity 5 tons, track gage 24 in.

Fig. 17.63—Steel frame Rock Car for Cia. de Cemento El Morro, Cuba. Capacity 4 cu. yds., track gage 30 in.

Fig. 17.64—Steel frame Army Car for Russian government, Russia. Capacity 5 tons, track gage 24 in.

Fig. 17.65—All-steel Hemp Fiber Car for Haytean American Development Corp., Hayti. Capacity 2½ tons, track gage 30 in.

Fig. 17.66 — Steel frame Flat Car for Central Venezuela, Venezuela. Capacity 15 tons; track gage 36 in.

Fig. 17.67 — Steel frame Flat Car for Gobierno Nacional de Colombia. Capacity 30 tons; track gage 4 ft. 8½ in.

Fig. 17.68—All-steel low-drop-side Gondola Car for Royal State Railways of Thailand. Capacity 27,550 lb., length inside 22 ft. 5¾ in.; track gage one meter.

Fig. 17.69—Four-wheel, all-steel Box Car for Royal State Railways of Thailand. Capacity 27,500 lb., length inside 21 ft. 6 in., track gage one meter.

Fig. 17.70—Steel frame Flat Car for Anglo Newfoundland Development Company, Newfoundland. Capacity 20 tons; track gage 42 in.

Fig. 17.71—Tank Car for Brazilian Government Railways. Capacity 5,000 gal. or 30 metric tons.

Fig. 17.72—High-side Gondola Car with hinged two-piece side doors for manual unloading. Belgo Minera Railways.

American Car and Foundry Export Company

Fig. 17.73—Tank Car for fuel oil, capacity 6,000 gallons.

Fig. 17.74—Fuel oil Tank Car for United Fruit Company—Compania Agricola de Guatemala. Capacity 7,500 gal.; track gage 36 in.

Fig. 17.75—Tank Car for F. C. del Sur, Costa Rica. Capacity 7,500 gal.; track gage 42 in.

Magor Car Corporation

Section 19

Freight Car Operation

Inspection; Interchange; Loading

COINCIDENT WITH THE CONSTANTLY INCREASING complexity of American industry, the interrelations between extended business interests and the necessity for supplying the requirements of a large, widespread population, the efficient handling of freight transportation has become an interline problem of national scope. To facilitate the economical movement of traffic under these conditions, the railroads have for many years permitted car-load freight to be shipped from consignor to consignee without the expense and inconvenience of transshipment at junction points. In addition, many railroads have established through merchandise or package freight services for the expeditious handling of less-than-carload lots. This has resulted in a vast interchange of traffic between the various railroad systems and in the gradual formulation of restrictions or rules under which this traffic interchange is conducted.

CONTENTS OF SECTION 19

	Page
Code of Car Service Rules	1239
Code of Per Diem Rules	1239
Code of Interchange Rules	1240
Loading Rules	1240
Prices for Labor and Materials	1240
Car Inspection	1241
Freight Car Utilization	1242

The formulation and enforcement of rules governing car interchange are important functions of the Association of American Railroads. Through its divisions, sections and committees, this association, working in voluntary cooperation with the railroads, constitutes an administrative institution for the regulation and control of car interchange and car relocation, except insofar as the Interstate Commerce Commission exercises public supervision.

Various codes of rules have been developed by the Association. Among the more important are: The Code of Car Service Rules, the Code of Per Diem Rules and the Code of Interchange Rules. In subscribing to these codes the carriers sign formal agreements whereby they bind themselves to faithfully abide by all the rules and stipulations set forth therein; and also to abide by each and all decisions or interpretations of arbitration committees, duly appointed for the purpose of settling questions or disputes that may arise under the respective codes.

Code of Car Service Rules

This code was adopted by the American Railway Association in October, 1900. Previously, the interchange of freight equipment had been conducted under a general resolution—"To prevent the misuse and diversion of freight cars"—adopted in 1892. For a number of years the code of 1900 was merely recommended by the Association and while fairly well adhered to by many of the roads, it was, nevertheless, frequently disregarded, especially during periods of car shortage. In 1912, a Commission on Car Service was appointed for the purpose of bringing about a more complete enforcement of the Car Service Rules. The functions of this commision were gradually increased from year to year up to 1920, when it was finally authorized to "Transfer cars from one railroad or territory to another when necessary to meet traffic conditions with due regard to car ownership and requirements."

The Car Service Division of the Association of American Railroads now supersedes this commission and during record-breaking traffic movements, the distribution and interchange of freight equipment has been handled with outstanding success.

The Code of Car Service Rules are principally concerned with the use and home routing of foreign cars. They provide for the automatic return of interchanged equipment under the various conditions of traffic movement and are to some extent interrelated with the Code of Per Diem Rules whereby a basis of settlement is definitely established between railroad companies handling freight equipment in interchange.

Code of Per Diem Rules

The per diem basis of settlement was first adopted in April, 1902, and became effective July 1, 1902, with 20 cents per car per day as the established rate. Prior to the year 1902, settlement between railroad companies for the use of foreign equipment was effected on a mileage basis. Since 1902 the per diem rates have been gradually increased until at present $1.00 per day is the established charge with the addition of varying penalties that come into effect under certain conditions covered by Rule 1, as follows:

"The rate for the use of freight cars shall be $1.00 per car per day, which shall be paid for every calendar day and shall be known as the per diem rate; except that when per diem is not reported to car owner within four months from the last day of the month in which it is earned, the rate shall be increased fifteen (15) cents per car per day for each six months' period or fraction thereof that report of such per diem is thereafter withheld; provided that the aggregate increase in the rate shall not exceed 60 cents per car per day."

The code contains twenty rules which provide in detail for all important contingencies affecting settlements between railroad companies for cars used in interchange; it prescribes methods for the making of interchange, junction and monthly per diem reports and formulates a definite arbitration procedure under which the per diem rules are interpreted and all questions or disputes pertaining thereto are finally decided. This arbitration procedure is administered by the Per Diem Rules Arbitration Committee appointed by the Transportation Division of the Association of American Railroads.

Code of Interchange Rules

Prior to the year 1864, there was little interchange of cars between the various railroad systems. Gradually, however, roads having the same gage began to interchange each other's equipment to avoid the inconvenience and expense incidental to breaking bulk and transferring freight at junction points. It soon became evident that rules of some kind were absolutely necessary if this interchange was to be handled in an equitable manner and with due regard to the rights of the various owners. This was especially evident in connection with the repair of foreign cars, settlements for damages, billing charges and similar matters. Consequently, at a meeting of the Master Car Builders' Association in December, 1876, a code of rules covering freight car interchange was adopted. The original code was brief and provided merely for the more important necessities. Since 1876, however, these rules have been greatly increased and extended until at present they function in such a manner as to govern the condition of, and repairs to, freight cars for the interchange of traffic in all its phases and complexities. Their purpose and intent may, perhaps, be best indicated by quoting the preface as follows:

These rules are formulated as a guide to the fair and proper adjustment of all questions arising between car owner and handling company with the intent of:

1. Making car owners responsible for and, therefore, chargeable with the repairs to their cars necessitated by ordinary wear and tear in fair service; by the Safety Requirements and by the Standards of the American Railway Association.
2. Placing responsibility with and providing a means of settlement for damage to any car, occurring through unfair usage or improper protection by the handling Company.
3. Providing an equitable basis for charging such repairs and damages.

Inspection of freight cars for interchange and method of loading will be in accordance with this code of Rules and the Specifications for Tank Cars and Loading Rules issued by this Association. Methods of recording inspection and repairs of cars and billing therefor under these Rules will also be governed by Supplementary Regulations issued in connection therewith.

Subsequent to the amalgamation of the Master Car Builders' organization with the American Railway Association, this code has been administered by the Mechanical Division of the Association and its rules are interpreted by an Arbitration Committee appointed by the General Committee of that Division. The Arbitration Committee consists of eight members, six representing railroads of the United States and Canada, and two representing private car owners. Five members of this committee constitute a quorum.

All questions or disputes arising under the rules of the Association of American Railroads governing the interchange of traffic between railroads subscribing to these rules, may be submitted to this committee for consideration and settlement.

Matters in dispute, to receive consideration by the Arbitration Committee, should be submitted in abstract form consisting of not more than three typewritten pages, letter size, single space, setting forth:

1. An agreed statement of facts.
2. Argument of plaintiff.
3. Argument of defendant.

The agreed statement of facts to be signed by both parties to the dispute.

Decisions of the Arbitration Committee are final and binding upon the parties concerned. These decisions are reported to the Association of American Railroads, Mechanical Division, and are incorporated in the annual proceedings of the association.

The two standing committees of the Mechanical Division of the Association of American Railroads who have to do with the interchange of cars and who report annually are: The "Arbitration Committee" who handle appeals under interchange rules and who consider and report on proposed changes to these rules, and the "Committee on Prices for Labor and Materials." The latter committee handles all matters having to do with the prices for labor and materials used in the repairs of foreign cars, as covered by rules 101, 107, 111 and 112 for freight cars and rules 21 and 22 for passenger train cars.

A book embodying the code of rules governing the condition of, and repairs to, freight and passenger cars for the interchange of traffic is issued annually, with supplements as required by the Association of American Railroads.

Loading Rules

The original rules governing the loading of lumber, logs, stone and other commodities were adopted by the former Master Car Builders' Association as recommended practice in 1896, and were advanced to standard in 1908. At present the further development and formulation of standard loading methods comes under the jurisdiction of a Committee on Loading Rules appointed by the Mechanical Division of the Association of American Railroads.

The proper loading, bracing and blocking of cars are generally conceded to be just as important in their bearing on safe, economical transportation as is the condition of the cars themselves and their various appurtenances. It is, therefore, essential that shippers, inspectors, mechanical supervisors and others concerned with the movement of freight traffic, should be thoroughly informed as to the best-loading methods in order that accidents caused by defective loading shall be eliminated, and damage claims reduced to a minimum. Loading rules are formulated by the above committee with the help and advice of many large shippers. The suggested methods are carefully studied; their practicability is demonstrated by numerous trial shipments before they are finally adopted after submission by letter ballot to the members of the Association.

Reports of the "Committee on Loading Rules" are submitted annually and following revision made by letter ballot, a book or supplement is issued embodying the rules governing the loading of lumber, logs, stone, etc., and the loading and carrying structural materials, plates, rails, girders, etc.

The transportation of explosives is governed by the regulation of the Bureau of Explosives.

Prices for Labor and Materials

The prices that are charged for labor and materials in repairing cars of foreign roads in interchange traffic is governed by a fixed scale. See the annual reports of the Standing Committee on this subject.

Car Inspection

On the railroads of this country, the complete and thorough inspection of all equipment used in traffic movement constitutes a very essential and important detail. Ordinarily freight cars are inspected at receiving yards, at classification yards, at loading points and on the various repair or shop tracks. Cars received from connections are inspected to make certain that this equipment conforms in detail with the Association of American Railroads' Rules of Interchange and to protect the receiving line from improper charges for delivering line defects. Cars offered in interchange, designated for certain classes of lading are also inspected to see if they are fit for such lading.

In classification yards cars are inspected to insure that trains will be in first-class condition before they leave the terminal. This involves the inspection and testing of air brakes, the detection of possible defects developed during yard switching, together with a thorough examination of journal boxes, lubrication facilities and other details.

Car inspection at industrial establishments and other loading points is performed to see that all cars are in proper condition before they are allowed to move in traffic. This inspection is also made to insure the suitability of cars for the proposed lading and to see that the loading rules are adequately observed.

The inspection of cars on repair or shop tracks involves the careful and complete examination of all car details to determine the extent of repairs necessary, whether or not the cars are safe to move traffic after receiving light running repairs or whether a more extensive overhauling is demanded. The kind and quantity of material to be ordered, the disposition of working forces and the functioning of the entire car repair organization very largely depend on the results of these repair track inspections.

Car inspections, as briefly outlined above, are made under the jurisdiction of Car Department Foremen and the Chief Interchange Car Inspectors. These officials and others concerned with the inspection and repair of rolling stock, compose an organization entitled the Railway Car Department Officers' Association. This association is devoted to the study of car inspection and repair problems including the Rules of Interchange, Loading Rules, Safety Appliance requirements and all matters pertaining to the maintenance of railway car equipment in a safe and economically operative condition.

Rules for Examination of Car Inspectors.

A.A.R. Recommended Practice—ADOPTED, 1902.

(Manual, Page L-31)

REQUIREMENTS.

One year at oiling cars.

Two years at car repairing.

Age limit for new men, thirty years.

Age limit for promoted men, forty years.

Vision, 20-20 in one eye and not less than 20-40 in the other, without glasses.

Method of Testing

Acuity of Vision.—The test card should be hung in a good light and the party to be examined should, if possible, be seated with his back to the window. Each eye should be examined separately, using, for the purpose of excluding one eye, a folded handkerchief. The lowest line that can be read should be determined by exposing only one letter at a time through a hole cut in a strip of cardboard. In making out the report in each case, the visual acuity of each eye should be denoted by a fraction of which the numerator represents the number of feet at which the applicant is seated from the card, while the denominator represents the number of feet at which the lowest line which he can read should be read. Thus, if at 20 ft. he reads the line marked 20 ft., his vision—20-20 or 1, which is the normal standard. If at the same distance he only can read the line marked 70 ft., his vision—20-70. If at 20 ft. he reads the 15-ft. line, the vision—20-15, or more than normal. If a room 20 ft. long can not be used, a testing distance of 15 or 10 ft. should be employed, in which case normal vision would be represented by 15-15 or 10-10 respectively, and lower grades of vision by such fractions as 15-20, 10-70, and so on.

Field of Vision.—Test should be made by having the applicant and examiner stand about 3 ft. apart, each with one eye shut, looking each other steadily in the eye. The examiner should then bring his hand in from the edge of the field toward the center of the space between them, until the applicant sees it coming. This should be done from different directions, up, down and from each side. The applicant should see the hand coming about as soon as the examiner does. If not, this should be noted on the report.

Hearing.—Test should be made in a quiet room. First, the examiner should hold the watch opposite the ear to be examined not less than 48 in. distant, then gradually approach the ear until the applicant hears the tick, the stop being used to satisfy the examiner that the applicant is not deceiving. The distance at which the applicant hears the watch should be noted in inches. The normal ear should hear the tick of the watch at 48 in. Then the hearing power will be denoted by a fraction whose numerator represents the number of inches at which the watch is heard. Thus, if he hears the watch at 48 in., his hearing—48-48, or normal. If he hears it only at 10 inches distant, his hearing—10-48, and so on.

Color.—The committee does not think it essential that inspectors should be rejected on account of imperfect color sense. It is, however, believed that inspectors should be tested as to their color sense so that they, as well as their employer, may know their condition in this respect.

Education.—The applicant should be able to write a legible hand in English, and also to read manuscript matter as well as printed matter.

Car Knowledge.—The inspectors should be able to name each part of the cars in general use, in preference using Car Builders' dictionary terms.

A.A.R. Rules.—Inspectors must pass a satisfactory examination on A.A.R. Interchange and Loading Rules, answering 75 per cent of the questions submitted. These questions should be of about the following character:

1. What are the A.A.R. Interchange Rules?
2. What is the object of the A.A.R. Interchange Rules?
3. What is the underlying idea or principle of these rules?
4. When is a company, operating the cars of another company, responsible for defects of such cars?
5. When a company is thus responsible, what should it do?
6. What care should be given to foreign cars by the company hauling them.
7. What cars must be accepted in interchange?
8. What is a defect card and how is it used?

9. Under what conditions is a road obliged to accept a car which is carded for defects for which the owner is not responsible?

10. What are the defects of wheels and axles for which owners and delivering companies are responsible?

11. Describe the form and use of the A.A.R. wheel gage?

12. What are the rules which apply to the cleaning of triple valves and cylinders?

13. What does the limit of height of drawbars mean?

14. When a company is obliged to make improper repairs, what must it do to call attention to such repairs?

15. What does the term unfair usage mean?

16. What are the rules regarding splicing sills?

17. What is the purpose of the repair card?

18. How do these rules apply to switching roads?

19. Are switching roads allowed to render bills against owners direct for repairs of any other than those named in Section 23 of Rule 5?*

**Note*—Refers to an old Edition of Interchange Code of Rules.

Note—Reports and discussion on the subject of apprentices will be found in the Proceedings of the Mechanical Division of the A.A.R. for 1924, 1925 and 1929.

Additional valuable information will be found in the paper on "Uniform Method of Selection, Education and Examination of Car Inspectors" and the discussion thereon contained in the 1929 Proceedings of the Master Car Builders and Supervisor's Association.

Freight Car Utilization

In 1943 the railroads of the United States are handling the largest volume of freight traffic in their history. Moreover, this is being done in spite of a heavy decrease in recent years in locomotive and car inventories. Some of the factors making these results possible are (1) Improved distribution of freight cars due to the efforts of the Car Service Division of the Association of American Railroads. (2) Efforts of the Mechanical Departments in maintaining the condition of locomotives and freight cars. (3) Increased loadings per car. (4) Heavier trains and higher train speeds.

Carloadings. For many years the weekly reports of carloadings have been carefully watched and used as a measure of the country's total industrial activity and production and as a measure of railroad performance. These statistics were a fairly satisfactory index as long as the average load per car and the average distance freight must be hauled remained constant. However, the changes in freight car loadings and in increased distance hauled in recent years has so completely altered the set up that "carloadings" are now considered as of secondary consideration and may even be misleading. In fact recent increases in traffic have been accompanied by decreases in carloadings. As an example, carloadings in 1929 totaled 52,827,925 cars, this fell to 28,179,952 cars in 1932 and was only 42,818,739 cars in 1942.

Revenue Ton Miles. These figures are now considered as a far more accurate measure of industrial production and railroad performance than carloadings.

Average Load for Freight Car. In 1927 the average load per freight car, including less than carload lot freight as well as carload freight, was 27.2 tons. This fell to 24.9 tons in 1932 and rose to 31.5 tons in 1942. Considering only carload freight the average load per car in 1942 was 40.1 tons, the first time on record where a load of 40 tons was exceeded.

Capacity per Car. In 1930 the average capacity of a box car was 41.5 tons and in 1942 it was 45.5 tons. Taking all revenue freight cars the average capacity was 46.6 tons in 1930 and 50.5 tons in 1942.

Freight Train Load. The average net tons per train was 676 tons in 1922, 663 tons in 1932 and 1,030 tons in 1942.

Freight Cars per Train. In 1921 the number of freight cars per train, exclusive of caboose, was 37.4 and in 1939 the average was 48.5 cars.

Car Miles per Freight Car Day. Taking all freight cars this was 50.1 miles in 1943.

Sources of Information. For information on freight car performance the following should be considered.

I. C. C. Statistics of Railroads in the United States (Annual).

I. C. C. Bureau of Transport Economics and Statistics (Monthly).

Statistics of Railways—Association of American Railroads, Bureau of Railway Economics.

Railway Age—Annual Statistical Issue (January) and Freight Progress Annual (May).